New Waves in Philosophy

Series Editors: **Vincent F. Hendricks** and **Duncan Pritchard**

Titles include:

Otavio Bueno and Oystein Linnebo (*editors*)
NEW WAVES IN PHILOSOPHY OF MATHEMATICS

Boudewijn DeBruin and Christopher F. Zurn (*editors*)
NEW WAVES IN POLITICAL PHILOSOPHY

Vincent F. Hendricks and Duncan Pritchard (*editors*)
NEW WAVES IN EPISTEMOLOGY

P. D. Magnus and Jacob Busch (*editors*)
NEW WAVES IN PHILOSOPHY OF SCIENCE

Yujin Nagasawa and Erik J. Wielenberg (*editors*)
NEW WAVES IN PHILOSOPHY OF RELIGION

Jan Kyrre Berg Olsen, Evan Selinger and Søren Riis (*editors*)
NEW WAVES IN PHILOSOPHY OF TECHNOLOGY

Thomas S. Petersen, Jesper Ryberg and Clark Wolf (*editors*)
NEW WAVES IN APPLIED ETHICS

Sarah Sawyer (*editor*)
NEW WAVES IN PHILOSOPHY OF LANGUAGE

Kathleen Stock and Katherine Thomson-Jones (*editors*)
NEW WAVES IN AESTHETICS

Forthcoming:

Jesús Aguilar, Andrei A. Buckareff and Keith Frankish (*editors*)
NEW WAVES IN PHILOSOPHY OF ACTION

Thom Brooks (*editor*)
NEW WAVES IN ETHICS

Allan Hazlett (*editor*)
NEW WAVES IN METAPHYSICS

Nikolaj Pedersen and Cory Wright (*editors*)
NEW WAVES IN TRUTH

Future volumes

New Waves in Philosophy of Mind
New Waves in Meta-Ethics
New Waves in Formal Philosophy
New Waves in Philosophy of Law

New Waves in Philosophy
Series Standing Order ISBN 978–0–230–53797–2 (hardcover)
Series Standing Order ISBN 978–0–230–53798–9 (paperback)
(outside North America only)

You can receive future titles in this series as they are published by placing a standing order. Please contact your bookseller or, in case of difficulty, write to us at the address below with your name and address, the title of the series and the ISBN quoted above.

Customer Services Department, Macmillan Distribution Ltd, Houndmills, Basingstoke, Hampshire RG21 6XS, England

New Waves in Philosophy of Science

Edited by

P. D. Magnus
University of Albany, USA

and

Jacob Busch
University of St Andrews, UK

palgrave
macmillan

First published 2010 by
PALGRAVE MACMILLAN

Palgrave Macmillan in the UK is an imprint of Macmillan Publishers Limited, registered in England, company number 785998, of Houndmills, Basingstoke, Hampshire RG21 6XS.

Palgrave Macmillan in the US is a division of St Martin's Press LLC, 175 Fifth Avenue, New York, NY 10010.

Palgrave Macmillan is the global academic imprint of the above companies and has companies and representatives throughout the world.

Palgrave® and Macmillan® are registered trademarks in the United States, the United Kingdom, Europe and other countries

ISBN 978–0–230–22263–2 hardback
ISBN 978–0–230–22264–9 paperback

This book is printed on paper suitable for recycling and made from fully managed and sustained forest sources. Logging, pulping and manufacturing processes are expected to conform to the environmental regulations of the country of origin.

A catalogue record for this book is available from the British Library.

A catalog record for this book is available from the Library of Congress.

10 9 8 7 6 5 4 3 2 1
19 18 17 16 15 14 13 12 11 10

Printed and bound in Great Britain by
CPI Antony Rowe, Chippenham and Eastbourne

Contents

List of Figures

Series Editors' Preface

New Waves in Philosophy Series

The aim of this series is to gather the young and up-and-coming scholars in philosophy to give their view of the subject now and in the years to come, and to serve a documentary purpose, i.e. 'this is what they said then, and this is what happened'. It will also provide a snapshot of cutting-edge research that will be of vital interest to researchers and students working in all subject areas of philosophy.

The goal of the series is to have a New Waves volume in every one of the main areas of philosophy. We would like to thank Palgrave Macmillan for taking on this project in particular, and the entire *New Waves in Philosophy* series in general.

Vincent F. Hendricks and Duncan Pritchard
Editors

Notes on Contributors

Justin Biddle is Assistant Professor in the School of Public Policy, Georgia Institute of Technology, USA. He received his PhD in the History and Philosophy of Science from the University of Notre Dame in 2006. His primary research interests are in the philosophy of science, social epistemology, and the ethics of science. He has published in journals such as *Social Epistemology*.

Otávio Bueno is a Professor at the Department of Philosophy, University of Miami, USA, before which he was an Assistant Professor at California State University and at the University of South Carolina where he was also made Associate Professor. He received his PhD from the University of Leeds in 1999. He has done extensive work in the area of Philosophy of Mathematics as well as in Philosophy of Science and has published on these issues in journals such as *Synthese, Erkenntnis, Studies in the History and Philosophy of Science, Philosophy of Science* and many more. Amongst his many other achievements he is co-editor of the volume *New Waves in Philosophy of Mathematics*.

Jacob Busch (PhD University of Auckland, 2006) is a research fellow in philosophy at the University of St Andrews and associate research fellow with the NAMICONA Research Centre at the University of Aarhus. Before moving to St Andrews he was a visiting researcher with the HPS, at Leeds (2005–2006). His work has appeared in journals such as *International Studies in the Philosophy of Science* and *Studies in History and Philosophy of Science*. He specializes in the metaphysical aspect of the realist/anti-realist debate in the philosophy of science.

Anjan Chakravartty is Associate Professor, University of Toronto Institute for the History and Philosophy of Science and Technology, and Department of Philosophy, Canada. He received a BSc in Biophysics from the University of Toronto and a PhD from the University of Cambridge in 2001. From 2000–2002 he was a Research Fellow at King's College, University of Cambridge. He is the author of numerous articles published in journals such as *Philosophy of Science, Synthese, Studies in the History and Philosophy of Science*, and the book *A Metaphysics for Scientific Realism: Knowing the Unobservable* (2007).

Gregory Frost-Arnold is Assistant Professor at the University of Nevada, Las Vegas, USA. He received his PhD from the University of Pittsburgh in 2006. Dr Frost-Arnold has published in journals such as *Philosophy of Science, Biology and Philosophy*, and *The Journal of Philosophical Logic*.

Kristen Intemann is an Assistant Professor at Montana State University, USA. She received her PhD from the University of Washington in 2004. She specializes in the areas of values and science and feminist philosophy of science. Dr Intemann has published in journals such as *Philosophy of Science* and *Science and Education*.

P. D. Magnus (PhD University of California San Diego, 2003) is an Assistant Professor in Philosophy at the University at Albany, State University of New York. Before moving to Albany, he was a visiting assistant professor at Bowdoin College (2003–2004). His work in the philosophy of science has appeared in journals such as *Philosophy of Science, The British Journal for the Philosophy of Science, Synthese*, and *Social Studies of Science*. Much of his work has concerned the alleged underdetermination of theory.

Jay Odenbaugh is Assistant Professor at Lewis and Clark College, USA. Before that he was a lecturer at University of California, San Diego. He received his PhD from the University of Calgary in 2001. He specializes in Philosophy of Biology and Environmental Ethics. He has published in journals such as *Philosophy of Science, Biology and Philosophy* and *Environmental Values*. He is a contributor to the *Encyclopedia for the Philosophy of Science* edited by Jessica Pfeifer and Sahotra Sarkar (2006).

Laura Perini is Assistant Professor at the Department of Philosophy, Pomona College, USA. She has an MA in Biology from the University of California, Los Angeles. She received her PhD from the University of California, San Diego in 2002. She has been a fellow at the Center for Philosophy of Science, University of Pittsburgh and a visiting fellow at Dartmouth College Humanities Institute, and she previously taught at Virginia Tech. She has published in journals such as *Philosophy of Science* and *Philosophy and Biology* and contributed to the *The Encyclopedia of Philosophy of Science* edited by Jessica Pfeifer and Sahotra Sarkar (2006).

Jessica Pfeifer is Assistant Professor at the University of Maryland, USA. She completed her PhD in Philosophy/Science Studies at the University of California, San Diego in 1999. The primary focus of her research has been on the nature of modal notions, and in particular the natural necessity involved in lawful relations. More recently she has focused on questions in the Philosophy of Biology. Pfeifer is co-editor (along with Sahotra Sarkar) of the two-volume *The Encyclopedia of Philosophy of Science* (2006). She has also been a Visiting Fellow and is currently an Associate Fellow at the Pittsburgh Center for Philosophy of Science.

Sherrilyn Roush is Associate Professor at the University of Berkeley, USA. Before that she was an assistant professor at Rice University. She has a BSc in Mathematics and she received her PhD from Harvard University in 1999. Roush has published in a number of journals, such as *Journal of Philosophy, Synthese, Philosophy of Science* and *Studies in History and Philosophy of Modern Physics*. She is also the author of the book *Tracking Truth: Knowledge, Evidence and Science* (2005).

Juha Saatsi has an MSc degree in Theoretical Physics from the University of Jyväskylä, Finland and completed his PhD in Philosophy at the University of Leeds, UK in 2006. From 2005–2006 he was a postdoctoral fellow with the University of Manchester and from 2006–2007 a postdoctoral fellow with the University of Leeds. He is currently a teaching fellow at the University of Leeds. Dr Saatsi is the author of numerous articles in such journals as *Philosophy of Science, British Journal for the Philosophy of Science,* and *Studies in History and Philosophy of Science*. He is also co-editor of *The Structural Foundations of Quantum Gravity* (2006).

Daniel Steel is Associate Professor at Michigan State University, USA. He has a BA in Anthropology and received his PhD from the University of Pittsburgh in 2002. He held a fellowship at the Center for Philosophy of Science at the University of Pittsburgh 2005. He specializes in Philosophy of Social Science as well as general Philosophy of Science and Philosophy of Biology. He has published in such journals as *British Journal for the Philosophy of Science, Synthese, Philosophy of Science,* and *Philosophy of the Social Sciences*. He is author of the forthcoming book *Across the Boundaries: Extrapolation and Causality in the Biological and Social Sciences*.

Michael Weisberg is Assistant Professor at the University of Pennsylvania, USA. Since 2006 he has been an associate at the Center for Philosophy of Science, University of Pittsburgh. He has a BSc in Chemistry from University of California, San Diego and he received his PhD from Leland Stanford Junior University in 2003. He has published in journals such as *Philosophy of Science, Studies in the History and Philosophy of Science,* and *British Journal for the Philosophy of Science.*

Eric Winsberg is Associate Professor at the University of South Florida, USA. He received his PhD from Indiana University in 1999. He held a postdoctoral fellowship at Northwestern University from 1999–2001 and has been a fellow at the Center for Interdisciplinary Studies, University of Bielefeld from 2006–2007. Since 2008 he has been a fellow at the Institute of Advanced Studies, University of Durham. He has published in journals such as *Synthese, Philosophy of Science,* and *Journal of Philosophy.* He is currently writing a new book: *Science in the Age of Computer Simulation.*

Introduction

P. D. Magnus and Jacob Busch

The explicit aim of volumes in this series is to collect contributions from young researchers likely to dominate the discipline; for this volume, the discipline in question is philosophy of science. It has been our privilege to edit such an audacious project, but it has also been a great challenge. We can only make educated guesses about the future. We cannot say with certainty which recent topics will be central to the discipline. Even selecting among areas which will probably remain central, there are competing desiderata: for example, to balance perennial topics against others which have more recently attracted the attention of philosophers.

A number of subjects have recurred in debates among philosophers of science: induction and the justificatory status of ampliative modes of inference; the role of explanation and its metaphysical status; causation; the problem of demarcation; the role of metaphysics in science; probability theory and its influence on formal approaches to confirmation; and scientific realism against various kinds of anti-realism. One might also add issues arising out of twentieth-century physics, such as the interpretation of quantum mechanics and the nature of spacetime theories. And it is no longer the case that a philosopher can claim to account for science when only really accounting for physics. So one might add issues that arise in biology, chemistry, climatology, psychology, sociology, and other sciences.

Philosophy of physics, philosophy of biology, and others have become autonomous specialties in their own right. There are conferences and journals dedicated to each, and we could without difficulty have collected papers for a volume exploring the 'new waves' in just one of them. So we are the first to admit that any volume covering philosophy of science *tout court* will leave out a great deal which might reasonably have been included.

We did not pick a roster of topics and invite each author to report on the state of the art in one area. That might have resulted in a stultifying collection of field reports. Instead, we invited contributors who are doing exciting work in diverse areas of philosophy of science. We explained the aim of the volume. And then we let them tell us what needed to be written, which debates are the important ones, and how those debates might best be resolved.

A consequence of our approach is that the table of contents is not an exhaustive roster of topics. Also, unfortunately, not all of the invited authors were ultimately able to contribute. So topics like causation, the role of experiment, and formal confirmation are regrettably absent. The field is large enough, however, that some omissions were inevitable. We apologize for them, all the same.

The alternative to our approach would have meant filling out the contents of the volume as if it were a shopping list, with topics that could be checked off as they were put in the basket. That would make sense if philosophy of science were nothing more than a mismatched agglomeration of all of these specialties. On that agglomeration conception, the authors in this volume are working in disparate fields. Yet, once asked to identify a pressing issue – something they are working on now that will still be worth thinking about a decade from now – none selected boutique topics. All of the topics addressed in this volume are within the purview of general philosophy of science, rather than being of interest only to sub-specialists within a fragment of a specialty. The perennial issues of realism and representation are presented here in their twenty-first-century incarnations. Where specific examples are discussed, they are used to illuminate more general questions. For example: electron micrographs raise questions about the nature of visual representation. Environmental science and climate change are test cases for accounts of the relation between values and science. A central issue in the philosophy of social science ultimately turns on our conception of science itself. And so on.

For the purpose of our introduction, we'll treat the topics in five broad groups: (1) scientific realism, (2) science and metaphysics, (3) representation, (4) the relation between science and values, and (5) science and social structure.

1. Realism

Scientific realism is a stock issue in the philosophy of science, motivated in its current form by the failure of logical positivism in the middle of the last century. The subject has both benefited from and to some extent suffered from the amount of literature on the various aspects of realism. Today there is little agreement on what constitutes the most significant aspect of the debate, but epistemic concerns are inescapable; realists must rely on ampliative modes of inference.

Juha Saatsi maps out the territory, proposing that various ampliative strategies are best organized by the conceptual distinction between content-driven and form-driven ampliative inferences. In Saatsi's view, this distinction does more than just systematize the realism debate; it also makes apparent that the more form-driven arguments are seriously problematic. As a consequence, Saatsi suggests that realists should argue for what he calls 'rather content-driven' arguments. He sets out what he believes to be a challenge for any realist, namely to settle how best to construe these content-driven arguments in *general* terms.

Sherrilyn Roush focuses on the pessimistic meta-induction, which argues from the fact that past scientific theories have been proven false to the conclusion that our current theories will be proven false also. Roush suggests a novel solution to this argument, and argues that there is in fact no formulation of the argument that takes into account that the pessimist must appeal not merely to the falsity of our predecessors' theories but to the unreliability of their ways of coming to their beliefs (as confirmed by their repeated false conclusions). But even granting their unreliability, Roush argues that nothing follows from this about whether we have a right to our confidence in our particular theories. In establishing this, the pessimist must show that the reasons for believing that our predecessors were unreliable must be the same as the reasons for believing that we are unreliable but this has not been shown, since the difference in methods between us and our predecessors undermines the legitimacy of the pessimist's induction. Therefore there is no motivation for the suggested pessimism about the prospects for realism about science. As such, the ground has been cleared for the further development of realist positions in science.

2. Science and metaphysics

Metaphysics has had a somewhat rogue status amongst philosophers of science and the role of metaphysics continues to be a point of contention. With the collapse of logical positivism, no one would insist that metaphysical talk is *nonsense*. Nevertheless, the same empiricist sensibilities that motivate anti-realism motivate humble ontological commitments – metaphysics could be rejected as mere speculation.

Anjan Chakravartty asks about the proper role for metaphysical debate in philosophy of science and suggests that there is no one right answer. He is, in effect, arguing against the assumption that the debate between realists and anti-realists can be resolved. He is skeptical of the idea that there *should* be a single correct answer to the question of how much metaphysical speculation is appropriate in understanding science. Choices of where to draw the line, of how much metaphysics to do, are rather a matter of convention. Importantly, Chakravartty does not take this to motivate an empiricist stance about science; he takes himself to be a realist. His point is just that we should not expect there to be an Archimedean point from which one side can persuade the other to abandon their wrongheaded ways.

Jessica Pfeifer argues directly against one kind of anti-metaphysical view: nominalism. She argues that even the most sophisticated variety of nominalism will be unable to account for our inductive practices. A strict nominalist approach to language would leave us with no way of explaining that we do make inductive inferences, regardless of whether we are justified in doing so or not. In this way she exploits a fact about our epistemic practices, without taking a stance on whether this practice is justified. If her argument is successful, it's a victory for non-nominalist positions.

3. Representation

The question of how theories map the world is closely related to realism (the accuracy of those maps) and metaphysics (what the features are that appear on the maps). In the logical empiricist tradition, a theory was treated as a set of statements in a formal language. The question of how a theory represents was just the question of how language represents. Yet the statement view ultimately collapsed, and representation became a thornier problem. Chapters in this section address the general problem of scientific representation, the problem of determining when two representations are genuinely distinct (rather than just expressed differently), and the problem of visual representation.

Otávio Bueno sketches a framework for accommodating different features of scientific representation. Bueno views representation in scientific practice as being specific to contexts where only partial information is available. From this diversity we obtain partial isomorphisms between our theories and the objects represented. The framework promises to accommodate the diversity of representational mediums used in scientific practice, from models and templates to micrographs, and the output of various instruments.

Gregory Frost-Arnold and P. D. Magnus address the familiar problem of the underdetermination of theory by data – evidence is insufficient for establishing which of two different theories is correct. A case of would-be underdetermination could be averted if the alternatives were not *different* theories at all, but merely different formulations of the same underlying theory. They call this the identical rivals response, and argue that it involves reducing the ontological commitments of each formulation. They argue further that there is no logical criterion for determining whether this ontological withdrawal is appropriate. Instead, applying the identical rivals response (or not) is a strategic choice.

Laura Perini asks how visual images function as scientific representations. Since scientists use visual representations in their arguments, she insists, understanding scientific reasoning requires making sense of pictures. To make sense of them, she offers a semiotic analysis of pictures and diagrams as symbolic. This analysis also helps clarify scientific representation more generally.

4. The relation between science and values

It is a bit of hoary wisdom that science is *value free*. Philosophy of science is concerned with epistemic matters, and questions of value should be left to the ethicists. The chapters in this section explode this old conception.

Jay Odenbaugh considers what lessons general philosophy of science can learn by considering the environmental sciences. He focuses on three areas. First, he considers how environmental scientists rely on models and simulations which they know are inaccurate in important respects and uses this to inform our understanding of idealization in science generally. (This overlaps with the concerns about representation in the previous chapters.) Second, he considers the way value commitments affect environmental science and argues that it does not compromise scientific objectivity. Third, he considers the role that climate science plays in political debates. This role, he argues, raises questions about the function of controversy and consensus in science.

Justin Biddle and Eric Winsberg argue that, in the area of contemporary climate modeling, scientists cannot estimate the uncertainties of climate predictions in a way that is free from 'non-epistemic' considerations. To put the point differently, they argue that developments within climate modeling are influenced by values and choices. To be clear, they are not arguing that consensus regarding the causal connection between fossil fuel emissions and global climate change is problematic. Nor are they arguing for skepticism about the climate models that have been employed to reach this result. Rather, if they are right, the subtle influence of values in good climate science indicates that good science can be influenced by values.

Kristen Intemann's primary focus is feminist philosophy of science. Although concerned with how gender influences science, feminist work is more broadly concerned with how understanding this influence can inform our understanding of science more generally. For example, most feminist philosophers of science have held that social and political values can play a legitimate role in good science. Intemann distinguishes two positions: feminist empiricism and standpoint feminism. She argues that debates between these two positions are ultimately misguided. They are not as different as their debates might make them seem. Moreover, Intemann insists, a more plausible position can be developed by taking parts of each.

5. Science and social structure

Traditional philosophy of science has sometimes ignored the fact that science is a social activity. And it has also often taken the natural sciences, especially physics, to be paradigmatic of science. Social sciences were faced with two unsavory options: employ the methods of the natural sciences (naturalism) or articulate an alternative (interpretivism).

Daniel Steel argues that we should not accept either option in their traditional form. The methods of natural science cannot be characterized in a precise enough way to apply to both fundamental particles and societies. If we understand naturalism as the position that the natural and social sciences employ the same methods, then it is a non-starter. Yet this does not mean that interpretivists win all the traditional debates. Instead, Steel argues, we can see naturalism as a commitment to what he calls the Enlightenment ideal of a science. According to this ideal, social science aims to discover the causes and effects of social phenomena so as to inform social policy. Framing the debate in this way makes better sense in terms of general philosophy of science, and also promises to redirect the debate in more productive directions.

Michael Weisberg is concerned about science as a social activity, rather than with social science. He asks how the arrangement of scientists' cognitive labor affects the development of science – that is, how scientific communities might best be organized so as to facilitate progress and discovery. He discusses three areas of recent research: the *marginal contribution/reward* approach, the *epistemic networks* approach, and his own *epistemic landscape* approach. He discusses some specific work in the latter approach, modeling the scientific community as a mixture of followers and mavericks. He concludes by suggesting ways in which the approaches might inform one another.

Weisberg's chapter exemplifies our problem in trying to write a précis for this volume. The work he discusses has things to teach us both about specific questions and about science generally. Yet its importance for philosophy of science is not yet settled. And so it is with all the essays. In a decade or more, we may look back to see where the new waves crashed upon the old shore. But now we can only guess where each will end up.

We offer the present volume as a collection of things that seem important in the present philosophy of science, as judged by these authors.

1
Form-driven vs. Content-driven Arguments for Realism

Juha Saatsi[1]

I offer a meta-level analysis of realist arguments for the reliability of ampliative reasoning about the unobservable. We can distinguish *form-driven* and *content-driven* arguments for realism: form-driven arguments appeal to the form of inductive inferences, whilst content-driven arguments appeal to their specific content. After regimenting the realism debate in these terms, I will argue that the content-driven arguments are preferable. Along the way I will discuss how my analysis relates to John Norton's recent, more general thesis that the grounds for licit induction are always material.

1. Introduction

Scientific realists maintain that science progresses by making good inductive inferences about the unobservable world. Anti-realists, on the other hand, are skeptical about our inductive powers. Most of the literature either advocates or criticizes some particular realist argument. Here I take a much broader, meta-level perspective on a wide selection of realist arguments. The aim of this is two-fold: to impose a degree of order on the debate to begin with, and to evaluate the profitability or otherwise of certain *kinds* of arguments *en masse*.

More specifically, I will be concerned with the following two issues. Firstly, I wish to compare and evaluate a broad array of realist arguments by discerning their basic intuitions. I will focus, in particular, on the way in which the different arguments are underwritten by contrasting general views on induction. Secondly, I will consider the following question. Given our analysis of different realist arguments, how should realism be defended against the selective inductive skeptic (whose skepticism concerns the unobservable, not induction in general)? I will argue that the content-driven realist arguments are preferable, and that this can fruitfully focus and reform the realist project.

The chapter proceeds as follows. I will begin by looking at an array of realist arguments, first in the abstract (section 2), and then in terms of some existent arguments (section 3). Then in section 4 I will attend to some interesting parallels between my analysis of realist arguments and Norton's recent 'material' theory of induction. In the end (section 5) I will argue that the content-driven arguments for realism are preferable.

2. The realism debate: initial observations

The arguments considered below aim to defend realism against *selective* inductive skepticism, not against Hume's all-embracing skepticism.[2] We begin modestly with the premise that we have got some substantive inductive knowledge of unencountered observable affairs (the sun rising tomorrow, Jupiter having moons etc.). Assuming this much, the challenge is to argue that we should also take seriously some ampliative inferences to the unobservable.

The first task is to understand and sort out the various intuitions and motivations behind the different realist responses to this challenge. There's a broad array of realist arguments littered around the vast literature. Let us begin with some initial observations about these arguments.

1. The following schema captures the basic form of most realist arguments at some level of abstraction:

We are happy with *such-and-such* inductions $\{I_O\}$ about the observable.

Such-and-such inductions about the unobservable $\{I_U\}$ are *relevantly similar* to $\{I_O\}$.

We should be happy with $\{I_U\}$.

Given the nature of *selective* skepticism, realists naturally attempt to build a bridge between certain observable matters and certain unobservable matters by attending to the kinds of inferences that scientists make about these respective domains. In one way or another they argue that observability *per se* is epistemologically irrelevant in the face of the fact that certain inductive inferences are similar, so that countenancing one (class of) inference(s) whilst being skeptical about the other is unnatural or ad hoc. The above schema is, of course, only an abstract blueprint that underlies various concrete realist arguments. The actual arguments implementing this schema differ with respect to how inductive inferences are construed and classified, to spell out what the two classes of inductions $\{I_O\}$ and $\{I_U\}$ are, and in what sense they are 'relevantly similar'.

2. Different realist arguments are motivated by different understandings of the inductive method of science. Inference to the best explanation (IBE) is often the central point of contention. Some say it is a unifying feature of all scientific reasoning, and realism turns on arguing that explanatory virtues are truth-tracking virtues. The most optimistic line of thought appeals to IBE in a rule-circular fashion at the meta-level. Others avoid appealing to a meta-level IBE by arguing for realism directly on the grounds that scientific inferences can be construed as IBEs. Yet others argue for realism without appealing to explanation at all, preferring to leave it open whether any (extant) descriptive scheme captures the inductive method.

3. The aim and scope of realist arguments differ. Some wish to produce a single overarching argument that covers all 'mature' and 'successful' science, once and for all. Others are happy to produce one argument for realism about this and another for realism about that, in a more piecemeal fashion. We can talk about *wholesale* and *retail* arguments for realism-about-X, depending on the scope of the arguments (Magnus and Callender, 2004). A wholesale argument aims at justifying a realist attitude towards some large class of inferences $\{I_U\}$ for the purpose of bridging the alleged epistemological gap between the observable and the unobservable. A retail argument zooms in on some relatively small class of inferences $\{I_U\}$.

4. Realist arguments differ in their basic assumptions about exactly what aspect of scientific inductive inferences is germane to the realist project. Some arguments focus more on the *form* of these inductions, whilst others focus more on the *content*. We can talk about *form-driven* and *content-driven* realist arguments for realism-about-X, depending on whether the emphasis is on the form or the content, respectively. The form–content distinction is logically independent from the retail–wholesale distinction, but all paradigmatic wholesale arguments in the literature are form-driven, whilst retail arguments are content-driven.

The distinctions above are comparative and relative. These attributes are simply meant to serve to order the various realist arguments by discerning one argument as more form-driven (wholesale) or more content-driven (retail) than another. These abstract distinctions will become clearer with concrete examples, below.

Given these initial observations, we should now try to analyze the whole gamut of different arguments that have made their way into established realist literature. How do these arguments realize the basic argument schema in 1 above? Do their respective advocates differ in their attitudes towards induction in general? Why are some realists inclined to give wholesale/form-driven arguments, whilst others prefer retail/content-driven arguments? Can we analyze the pros and cons of these different leanings by abstracting away from the details of the particular arguments?

I will next take a closer look at the form–content distinction with reference to some specific realist arguments, and then compare the distinction to a recent view on induction advanced by John Norton, before coming back to the last question above.

3. The realism debate: an array of arguments

A realist argument is form-driven if it attempts to justify some inductive inferences by reference to some general formal attribute unifying all these inferences. Content-driven arguments, by contrast, take there to be more justificatory analysis to be done on case-by-case basis by taking into account what these inferences are about. The contrast between form and content-driven arguments admits of degrees. A set of inferences can share some single characteristic which in a realist argument becomes concurrently the vehicle of justification for each instance of inductive inference featuring that characteristic. Corresponding to the level of generality at which such characteristic is described – how encompassing the class of such inferences is – we have more and less form-driven realist strategies. This abstract preliminary distinction between form-driven and content-driven arguments is best clarified via concrete exemplars of actual realist arguments. I will consider the arguments in the table below.

Advocate	**Argument**	**Form vs. Content**
Boyd/Psillos	No-miracles argument	Extremely form-driven
Lipton	First-order explanationism	Rather form-driven
Hacking/Achinstein	Experimental realism	Rather content-driven
?	Scientific arguments	Fully content-driven

Extremely form-driven

The standard explanationist argument for realism, originating from Putnam, and finessed by Boyd and Psillos amongst others, is form-driven to the extreme. Psillos (1999) is a notable recent author in this lineage. He argues (roughly speaking) that the scientific method is based on inference to the best explanation, and that by meta-level use of IBE we can (in the framework of externalist epistemology) justify the scientific use of IBE as truth-tracking. Hence, we've got an attempt to justify realism by reference to an extremely general characteristic unifying all scientific inferences to unobservables. Namely, they are all of the same form: IBE.[3]

There are a couple of noteworthy ideas underlying Psillos's form-driven explanationism. First of all, Psillos (following Harman, 1965 and Josephson, 1998) takes IBE to be a fundamental, primitive foundational form of inductive inference (see Psillos, 2002, for example). Secondly, Psillos explicitly appeals to the similarity in the form of our reasoning about the observable and the unobservable matters, displaying clearly the underlying intuition:

> Theoretical beliefs in science are formed by means of abductive reasoning. But so are most of our every-day commonsense beliefs. Realists have exploited this fact in order to argue that if one has no reason to doubt commonsense abductive reasoning, then one should have no reason to doubt abduction in science. The *pattern of reasoning*, as well as justification, are the same in both cases. (Psillos, 1999: 211, my emphasis)

And on these grounds Psillos accuses the selective skeptic (van Fraassen) of adopting a selective attitude against inferences about the unobservable:

> Clearly, van Fraassen sustains a selective attitude towards IBE. The latter is a means of going beyond the realms of what has been actually observed and forming warranted beliefs about unobserved things and processes. Yet IBE is not [for van Fraassen] a means of forming warranted beliefs about the realm of unobservable things or processes. (Psillos, 1996: 34)

Rather form-driven

Many realists are wary of meta-level application of IBE, and also of the idea that there is pertinent justificatory unity to *all* scientific inferences that can be viewed as being abductive in form. Yet some of these realists wish to tap into the pivotal explanatory dimension of science, and appeal to explanatory virtues in a less form-driven way. These realists argue that the gap between ampliative inferences to observables, on the one hand,

and to unobservables, on the other hand, is bridged by virtue of the fact that the respective inferences are *not only* of the same general form (IBEs), *but also* of the same more specific 'inferential kind'.

Lipton (2004) presents such an argument. He develops an overarching *descriptive* account of confirmation and induction in terms of inference to the best explanation.[4] Regarding the justificatory challenge of realism, he puts forward a very general argument to unify and justify a significant class of abductive inferences of science. After repudiating the No-miracles argument, Lipton considers a less form-driven, first-order explanationist strategy:

> Can explanationism defend realism instead by appeal to the *structure* of those first-order inferences? ... The *structure of causal inferences* is the same, whether the cause is observable or not ... So there is a prima facie case for saying that all these inferences should be construed in the same way: granting the truth-tropism of inferences to observable causes, we ought also all to be realists about inferences to unobservable causes, since the *inferences have the same form* in both cases. (Lipton, 2004: 199–200, my emphasis)

Although Lipton avoids a form-driven abductive inference about science at the meta-level, he still provides a very general template for justifying scientific inferences. For him any scientific first-order instance of causal abduction is (probably) approximately true by virtue of being *formally similar* to everyday ampliative reasoning about the observable. Unlike Psillos, Lipton takes it to be incumbent on the realist to provide a more specific description of the kind of abductive reasoning that allows us to bridge the gap between everyday theorizing and scientific theorizing. Hence Lipton stresses the *causal-contrastive* mode of IBE. But is that enough said? I will return to this below (section 5).

How form-driven is this particular realist argument? It depends on how tightly the relevant 'inferential kind' is delineated. Just appealing to causal explanations (spelled out as contrastive explanations – cf. Lipton, 2004: ch. 3) yields a rather form-driven argument with very meager constraints on the content of the inference. The notion of contrastive causal explanation is a broad one, even at the level of observable matters.

Rather content-driven

At the more content-driven end of the spectrum there is a set of arguments whose advocates are collectively known as *experimental* realists.[5]

Experimental realists do not have to advocate any level of explanationism, not even as a significant descriptive thesis about the scientific method. Traditional questions about general characterization of induction are simply irrelevant to their realist arguments, for these arguments do not primarily rely on the idea that a scientific inference to some unobservable posit has a particular form. Rather, these content-driven arguments rely on case-specific considerations, typically closely following the reasons that scientists themselves supply for their beliefs about something unobservable. For example, regarding the paradigm unobservable entity, the atom, it has been popular to examine Perrin's original reasoning to the existence of atoms on the basis of Brownian motion (Achinstein, 2002; Miller, 1987; Salmon, 1984).

Although there are many content-driven realist arguments that differ considerably in detail and rhetoric, there is a common underlying pattern to be found. Each 'experimental' argument – I will question the aptness of the title below – is naturally construed as relying on some material assumption about the uniformity of the world that crosses the observable–unobservable boundary. The unwritten premise is that the uniformity assumption required for realism-about-X is as innocent as some assumption required by some induction about an unobserved observable. These uniformity assumptions pertain to particular matters of fact, and the epistemic warrant conferred by an experimental realist argument is correspondingly localized.

Achinstein (2002) is the latest (and the clearest) representative of this line of thought.[6] He analyzes Jean Perrin's reasoning to the existence of atoms as *causal-eliminative*, also giving a more general account of the conditions on which this kind of reasoning is justified. Achinstein (2002: 474) presents Perrin's reasoning as follows:

1. Given what is known, the possible causes of effect E (for example, Brownian motion) are $C, C_1, \ldots, C_n$ (for example, the motion of molecules, external vibrations, heat convection currents).
2. $C_1, \ldots, C_n$ do not cause E (since E continues when these factors are absent or altered).
   So probably
3. C causes E.

Observing the microscopic particles dancing around, continually accelerating and decelerating, indicates the existence of internal forces responsible for such behavior, assuming that no plausible external cause can be found. And the meticulous experiments performed by Guoy did indeed allow Perrin to eliminate the plausible external candidate causes $C_1, \ldots, C_n$. The various experiments performed by himself and others then allowed Perrin to claim quantitative evidence for his initial conclusion and for the numerical value of Avogadro's constant.

Such scientific reasoning immediately raises obvious anti-realist worries. First of all, there is the possibility that the hypothesis of internal molecular forces singled out by eliminative reasoning is merely the best of a bad lot. How do we know that all the possible alternative causes of the phenomenon have been cited and eliminated by experiments? Achinstein's response is to insist that the realm of possibility here is restricted by our background knowledge.

> The claim that the possible causes cited probably include the actual one can be defended by appeal to the fact that the *phenomenon in question is of a certain type* that, experience has shown, in other cases is caused by one or the other of the causes cited. (Achinstein, 2002: 478, my emphasis)

But this immediately raises a second anti-realist worry: how can we justify inferences to the unobservable on the basis of the observable, on the basis of what 'our experience has shown'? For example, in Perrin's argument we need to justify the inductive generalization from '*All observed* accelerating bodies in contact with other bodies exert forces on them' to '*All* accelerating bodies, including molecules (if any exist), in contact with other bodies exert forces on them' (ibid.: 481). And empiricists like van Fraassen, of course, take such inductive inferences to the unobservable to be unjustified and unjustifiable.

The way Achinstein responds to this second worry brings the content-driven character of his argument to the fore. According to Achinstein, the realist can provide a positive empirical reason for taking observability *not* to be a biasing condition for an inductive generalization from a sample.

> One can vary conditions or properties in virtue of which something is observable (or unobservable). For example, items can be observable (or unobservable) in virtue of their size, their distance from us in space or time, their duration, their interactions (or lack of them) with other items, and so on ... If we vary the conditions in virtue of which bodies are observable and find no differences in whether bodies have

> mass, and if we have no contrary empirical information, then we have offered an empirical argument to support the claim that the fact that all observed bodies are observable does not bias the observed sample with respect to the property of having mass. (ibid.: 484–485)

Arguably selective skepticism should feel serious tension here. The realist appeals to independence of mass with respect to variation in conditions and other properties that are relevant for observability, and arguably that sort of independence also counts for the legitimacy of certain ampliative inferences about unobserved observables. For example, having observed conservation of momentum for bodies of various sizes and masses we infer that conservation of momentum holds for bodies bigger and more massive than the ones we have so far observed. The selective skeptic is happy with this inference, but not with an inference that takes us from observed collisions to unobservable collisions. So whence the difference? After all, the logical possibility of *observability* being a biasing condition is on a par with the logical possibility of *having been observed* being a biasing condition. We can view the inductive inference to unobserved observables being underwritten by the specific assumption that

> (*) momentum conservation is independent of the size and mass of the colliding bodies, at least for all observable bodies.

The induction to unobservables is correspondingly underwritten by the assumption that

> (**) momentum conservation holds for all bodies regardless of their size and mass.

If we are justified in making some assumptions that unify the world beyond our observations, as the selective realist concurs, why would those assumptions *never* extend their reach to the unobservable? Isn't it simply ad hoc to happily go from our observations to (*), but deny that (**) is an illegitimate extrapolation? There doesn't seem to be anything in our experience to suggest an epistemological difference *in the content of these particular assumptions*.

Achinstein's argument is content-driven because it concerns only a very particular uniformity of the world: the relevant mass-related properties (e.g. conservation of momentum) are independent of the properties in virtue of which bodies are observable, or otherwise. Achinstein's realist analysis takes explicitly into account those particular matters of fact which underwrite Perrin's inductive argument for the existence of atoms. And his response against the selective skeptic turns on a kind of *tu quoque*: arguably our inductive inferences about the unobserved (but not unobservable) massive bodies are underwritten by assumptions that are epistemologically no different from the ones that the content-driven argument appeals to in this instance.

It seems that different realist arguments are motivated by different general views on induction. The advocates of form-driven arguments are inclined to look for an underlying unity in the form of scientific inferences to bridge from observables to unobservables, whilst the proponents of more content-driven arguments turn on our intuitions about some particular assumptions about the world. Many arguments of the latter kind are notoriously imprecise and rhetorical (but also intuitively pulling, e.g. Hacking, 1982), and a certain amount of interpretation is required. The interpretative gloss presented here, emphasizing the content-driven character of these arguments, gains impetus from the following observation. Many have reacted to these arguments by objecting that they are *just as abductive in form* as the form-driven realist argument based on the success of science (e.g. Resnik, 1994; Psillos, 1999). These reactions are misguided but unsurprising: if one is looking for a form-driven and formal justification of induction, and is furthermore tempted to massage any inductive argument into an inference to the best explanation, then the experimental realist arguments seem to rely on an abductive form. However, the intuition driving the content-driven arguments is that the justificatory work is done at the level of particular material assumptions. The fact that a realist argument turning on a particular uniformity assumption can be naturally construed as an inference to the best explanation is wholly irrelevant, given the very different view on what makes an induction licit. John Norton has contrasted these different perspectives on induction with his 'material theory of induction' to which I will now turn.

4. Parallels with Norton's analysis of induction

Norton (2003, 2005) provides a broad meta-level analysis of ampliative reasoning that supports a content-driven view of induction. In their attempt to describe good inductive reasoning, philosophers have identified many an inductive schema operative in the sciences and everyday life, from simple enumerative induction to fancy Bayesian logics. Norton calls such abstractions *formal* theories of induction: they attempt to provide a formal schema to distinguish licit inductive inferences independently of case-dependent detail. The basic motivation for Norton's content-driven view of induction comes from the realization that any extant formal schema furnishes both licit and illicit inductions, and therefore cannot by itself mark what makes an induction licit. Rather, a licit induction is licensed by local 'material postulates', not the form of induction *per se*. Hence, Norton advocates a *material theory of induction* according to which all licit inductions are ultimately 'underwritten by local material facts'.[7]

Consider the classic case of enumerative inductive generalization, for example. There are several good inductive arguments of this form, of course.

> Sample A of lead melts at 327.5 degrees Celsius.
> Sample B of lead melts at 327.5 degrees Celsius.
> Sample C of lead melts at 327.5 degrees Celsius.
>
> ———
>
> Any lead sample melts at 327.5 degrees Celsius.

This is (presumably) a licit induction. But this simple form of enumerative induction is, of course, illicit when the target of generalization is a kind which is less homogeneous with respect to the property in question. Hence, one cannot thus infer the melting point of any sample of plastic, say. What makes the inductive argument above licit is not its form, argues Norton, but the fact that it is an enumerative induction *about lead* (or about an element). This is typically left implicit, but it can be included explicitly as a premise – 'lead (as an element) is uniform in this respect' – rendering the argument (in this case) deductive, an instance of *demonstrative* induction. Norton calls such (typically unwritten) local premises 'material postulates'. A material postulate is a proposition stating a particular uniformity-fact.

Similarly, consider the following abductive argument, regarding the cause of solar eclipse.

> (OBS) The sun appears to be engulfed, little by little, by a massive object, leaving a halo.
> (BE) The best explanation of (OBS) is that the sunlight is obstructed by an intervening celestial body.
> (IBE) The best explanation is probably true.
>
> ———
>
> Therefore, the cause of the eclipse is an intervening celestial body.

Given suitable background knowledge this is a licit inference. What makes it licit? We should notice that there are two local parameters that need to be fixed (to get the premise (BE)) in order to apply the schema IBE. First of all, we need to say what *counts as an explanation* in this context. Why exactly is it that a bona fide explanation of (OBS) can be provided by a hypothesis about an intervening body, but not the hypothesis that the sun and the moon make love and discreetly hide themselves in darkness? We don't have, nor do we need to have, an overarching general theory of explanation to answer this. Rather, it suffices to look at the relevant background knowledge about the nature of light and the sun. Secondly, if we have different competing explanations, *how are these to be compared* so as to make (BE) true? What exactly is it that makes an explanation in terms of an intervening body *better* than, say, an explanation in terms of changing intrinsic nature of the sun? Again, such comparison turns on particular assumptions about the prevailing facts about the sun, instead of general facts about explanation. Hence, what makes the above argument licit is not the fact that it is *an instance of* abstract universal schema of inference to the best explanation. Rather, it is the fact that in this local context our explanatory judgements regarding the two parameters are such that they ensure this schema really functions. That is, our background-knowledge-dependent explanatory judgements reflect the relevant ways the world is, the *facts*. If we want to make explicit the material postulates that underwrite this inference, we need to write down the particular background assumptions that underlie (BE).

Enumerative induction and inference to the best explanation do not exhaust induction schemes, of course, but arguably the lesson generalizes (Norton, 2003). It is important to be clear here that Norton's thesis is first and foremost a *descriptive* one. It is one thing to describe good-as-opposed-to-bad inductive reasoning, say, and a whole other thing to *justify* some way of reasoning as profitable (truth-tracking, or empirical-adequacy-increasing, say) (Lipton, 2004). Norton's theory aims to locate a distinction between good inductions and bad inductions in the abstract, without making any further claim as to whether we are actually in a position to *know* which are which. It locates the distinction between licit and illicit not in the form (or any universally describable feature) of an inductive argument, but in its *content*. This philosophical theory about the distinction between good and bad inductive arguments does not in itself amount to knowledge that any particular inductive argument is good, since we may not know that the relevant local material postulates really represent *facts*. So justification is a further question.

Nevertheless, Norton does also suggest that his theory has interesting justificatory repercussions regarding Hume's problem of induction.[8] Hume's description of induction focused on enumerative generalization, and thus his argument against the possibility of justification of induction naturally turned on the idea that enumerative generalization hangs on the assumption of uniformity of nature. Although it is nowadays clear that enumerative induction is woefully inadequate as a description of our variegated ways of inductive reasoning, many think that Hume stated his argument in a general enough form for it to apply to any mode of non-deductive reasoning. Norton disagrees. For the way in which Hume's problem is typically presented relies explicitly on formal understanding of induction. Consider attempting an inductive justification of induction. We've got our first-order inductions about the world, and we've got a meta-induction about the past success of these first-order inductions. Such constructions, Hume's argument goes, are blatantly circular because both arguments are of the same *form*: 'more of the same'. If we are trying to thus establish the reliability of this formal schema of enumerative induction (irrespective of what the schema is applied to), we irrefutably end up running in circles. But according to the material theory of induction, no induction is licit purely by virtue of its form anyway. So the classic circularity predicament is based on a misconstruction of the whole justificatory challenge.

In the material theory of induction, by contrast, a good induction is grounded on the facts correctly described by the material postulate. So justifying a particular induction is a matter of justifying the relevant material postulate. This material postulate cannot be just taken as given, and justifying a particular material postulate requires another induction. But this is a *different induction*, grounded on *different facts* described by *different material postulates*. No circularity ensues, and arguably our best actual inductions are background-dependent and local in exactly this way:

> It merely describes the routine inductive explorations in science. Facts are inductively grounded in other facts; and those in yet other facts; and so on. As we trace back the justifications of justifications of inductions, we are simply engaged in the repeated exercise of displaying the reasons for why we believe this or that fact within our sciences. (Norton, 2003: 668)

This avoids the circularity problem, but isn't there an obvious regress here? Norton is optimistic in this regard:

> What remains an open question is exactly how the resulting chains (or, more likely, branching trees) will terminate and whether the terminations are troublesome. As long as that remains unclear, these considerations have failed to establish a serious problem in the material theory analogous to Hume's problem. (Norton, 2003: 668)

The problem at hand is different from Hume's, but it is potentially equally damaging. It seems that several everyday inductions yield beliefs which are justified on a par with our best scientific theories. Consider the paradigm induction of the sun rising tomorrow. Pre-scientifically such induction concerns some basic regularity of the world. We can now justify this regularity by appealing to different, scientific material postulates that represent more general facts about gravitation and dynamics. These facts are more general but they are still local by virtue of not being a priori universal postulates about worldly uniformity that equally applies to green emeralds, say. But can we take our best science-discovered regularities to be any better confirmed than these everyday inductions? And even if we have such reverence for scientific over everyday beliefs, it is not clear how the science-informed starting point – a more general basic regularity – is any less problematic *qua* basic regularity. Both regularities are inferred from a finite set of experiences.[9] Although there is a sense in which we have scientifically justified what was taken to be a primitive regularity beforehand – and hence accords with the scheme of material induction which '*describes* the routine inductive explorations in science' – the philosophical challenge of justifying induction concerns the respective starting points. What reason do we have to believe that our best science will still work tomorrow?

Despite these reservations, I fully subscribe to Norton's *descriptive* thesis about the locality of inductions and of their inherent background dependency. Regarding justification, it does show that one needs to be more careful about how the all-out justificatory challenge is posed. A typical two-line statement of *The* Problem of Induction is not in line with the fact that we do not have a universal formal schema (or a set of schemas) to capture the difference between licit and illicit inductions. However, an equally difficult problem of responding to a wholesale inductive skeptic may remain. But now we should return to the challenge posed by the selective skeptic, and I will argue that the content-driven view on induction does have ramifications regarding this more specific challenge.

5. Why content-driven realist arguments are preferable

Let's now return to questions posed at the end of section 2. Only one question remains: does the above analysis of realist arguments shed light on their pros and cons? The answer is 'yes'. Let's begin with the arguments that are driven by (more) emphasis on (more) formal similarities, whilst downplaying case-specific details. The seeming advantage of these arguments is that one gets more with less: a justification of a significant class of scientific inferences by their shared form, without having to pay much attention to what these inferences are about. This mirrors the appeal of form-driven descriptive theories of induction: being able to tell licit from illicit inductions by a mere formal schema. Norton has argued that the descriptive project is forced to take into account ever-present material assumptions. Similarly, I will presently argue, the justificatory project must pay heed to material assumptions. We will see that the downside of form-driven realist arguments is an increased epistemic risk that is unacceptable. By contrast, the (more) content-driven arguments have (more) emphasis on case-specific content. The advocates of content-driven arguments are happy to admit that justification of knowledge of the unobservable world is a business that always hangs on assumptions of particular matters of fact driving ampliative inferences. These arguments are epistemologically more secure, but the price to pay for this is the limited scope of each particular argument.

We can criticize form-driven realist arguments as follows. These arguments suffer from what could be termed the *Description–Justification Gap*. Too much emphasis is paid on formal descriptive unity, without realizing that descriptive unity can be cheap and does not amount to justificatory unity. Unified description can be achieved at the level of formal induction schemas simply by abstracting away from the specific material postulates that make particular inferences licit. But fully capturing what makes a particular induction licit requires more. Since a licit induction is always underwritten by a material postulate correctly representing a relevant fact, some justification must be given at that level as well. This is exactly what is missing in the realist arguments that attempt to cross the gap between the observable and the unobservable by comparing the respective inductive inferences vis-à-vis their form. In the extreme form-driven case, for example, the ambitious meta-level use of inference to the best explanation appeals to descriptive unity that spans from scientific to *philosophical* explanations.[10] Given how flexible the required unifying characterisation of IBE is, the two parameters (cf. section 4) determining what counts as an explanation, and what counts as a good explanation, are left wide open. But the mere form of an inference cannot carry the justificatory burden.

A realist espousing a form-driven argument may argue that in the *absence* of any reason to think that scientists' IBE-infested methodology becomes suddenly unreliable in some unobservable domain, our epistemic attitude to scientists' conclusions should be uniform regardless of whether the targets of investigation are observable or not. But this is insufficient, and more is required: a *positive* reason to think that the respective inferences are in same epistemological boat. Having a reason to suspect an inductive inference in a particular context requires a reason to suspect that the material postulate underwriting that inference does not correspond to worldly facts. For example, we have such grounds to suspect the inductive generalization from 'All Turkish adult males I've encountered are bearded' to 'All adult male Turks are bearded'. We know enough of *human beings* to know that nationality simply isn't a strong enough unifying factor in this respect. But *not* having such negative grounds for suspecting an induction does not amount to having positive grounds for it, either. For example, we may not be able to give any such *particular* reason to assume that inferring to the best explanation of the success of science is less likely to yield a true conclusion than inferring to the best explanation of a solar eclipse. But without any *positive* reason to think that the respective explanatory virtues are on a par as inductive virtues we are left wondering whether the material postulates underwriting the respective inductions really represent worldly facts. Why assume that the world is a place in which enquirers like us evolve so as to develop the ability to assess what is *really* the explanatory truth behind the success of science? Some answer needs to be given, and it is cold comfort if the advocate of the form-driven argument can respond only by pointing to the fact that the realist inference can be viewed as having the *form* of inference to the best explanation.

Similarly, we may not have any particular reason to suspect that scientists' appeal to explanatory virtues in quantum physics is any less reliable as an inductive guide than farmers' evaluation of the explanatory virtues required to catch a flock-harassing beast. But what we need is a positive reason to think that the respective explanatory virtues are on a par as inductive virtues. And just appealing to the abductive form of the respective inferences isn't enough on its own, given the huge difference in the two domains and the kinds of inferences made, even if the inferences do belong to a unified kind at some level of abstraction, when we leave it open enough what counts as an explanation and how explanatory goodness is measured.

The advocates of form-driven justificatory arguments are under pressure to shift emphasis on the contents of inductive inferences in order to rule out illegitimate use of cheap descriptive generalizations. For this reason Lipton (2004), for example, focuses more narrowly on *causal-contrastive* explanations: allegedly the epistemic status of an inference to the best causal-contrastive explanation is unrelated to the explanans being observable or otherwise. But the notion of causal explanation at stake is still rather open, and Lipton attempts to justify a very broad class of explanation-driven inferences by a single realist argument. Although the unifying characteristic here is not *purely* formal – a causal explanation obviously needs to reflect a *causal fact* about the world – it does not seem that Lipton's schema captures what makes each instance of causal-contrastive abductive inference licit. There is still much contextual variability in how the best explanation is chosen.

Lipton explains that 'for the causal explanations of events, explanatory contrasts select causes by means of the Difference condition: *To explain why P rather than Q, we must cite a causal difference between P and not-Q, consisting of a cause of P and the absence of a corresponding event in the case of not-Q*' (2004: 42).[11] This is a rather abstract characterization of what is required of good causal-contrastive explanations, and ultimately much hangs on case-dependent detail. For example, the notion of 'corresponding event' is highly contextual, and gets fixed by factual assumptions regarding the situation at hand. These material assumptions are needed to tell the licit uses of this schema from the illicit ones. We can leave these assumptions implicit when we deal with mundane explanations of observable matters by observable causes. But moving further away from everyday experience increases the risk of misjudging the explanatory status of causal hypothesis, and the mere form of a causal-contrastive IBE is not enough to warrant it. One can respond by further narrowing down the class of causal-abductive inferences by explicitly fixing more specific assumptions. This clearly yields a more content-driven realist argument. But how content-driven do we need to go?

6. Conclusion: the realism debate reformed

The justificatory challenge for the realist is to argue for a *Unity of Inductions* that makes selective skepticism unnatural and unappealing. The literature contains a great variety of ways to argue for such unity, and I have argued that the intuitions behind the different arguments correspond to more content-driven and more form-driven understanding of induction. This offers a useful way to order the sprawling debate for comparison and evaluation of the alternatives.

The conceptual distinction between form-driven vs. content-driven not only serves to systematize the realism debate; it also has deeper repercussions. For one thing, the more form-driven arguments are found to be seriously problematic. This is due to the gap between achieving a descriptive unity and achieving a justificatory unity: we haven't been given any positive reason to think that the kind of descriptive unity that the form-driven arguments capitalize on amounts to a relevant justificatory unity. This pushes the realist towards the more content-driven argumentative strategies. I have argued that the 'experimental' realist arguments can be viewed as (rather) content-driven arguments. But this raises further questions. What is the best way to construe these content-driven arguments in *general* terms? Exactly how content-driven are they? I'll finish the chapter with some tentative remarks on these issues.

The material theory of induction acknowledges that any licit inductive argument has both a form, and an underlying material postulate. The form-driven realist arguments argue for the Unity of Inductions at the level of shared form, whilst the content-driven arguments depend on an analysis of the relevant material postulates. Hence, the justificatory work in the latter is done by comparing the material postulates pertinent to some induction-to-observables, on the one hand, and some induction-to-unobservables, on the other. But this way of putting it makes it clear that these arguments have nothing to do with 'experiments' or 'entities' *per se*. Rather, they have to do with a content-driven comparison of the respective inductions.

The material postulates underwriting inductions to the observable and to the unobservable, respectively, are still going to be different, of course, so there is no question of *identifying* the postulates required by the realist with those required by the selective skeptic. The best one can do is still a judgement of *naturality*, or otherwise, of drawing the line of epistemic incredulity at a particular point. But this is how realists have always argued, admitting that there is always ample *logical* room for inductive skepticism, selective or not. This is just the nature of induction *qua* non-deduction. But of the various ways of arguing against the unnatural skepticism of the anti-realist, the content-driven approach, I maintain, is the best. For if the particular material facts are what make an induction licit, then a realist appealing to descriptive unity takes an unnecessary epistemic risk. Appealing to the form of an inference, instead of its material postulates, raises the possibility that an inductive inference is taken to be licit when there is no relevant material fact to underwrite it. Of course, the more abstract the unifying description, the higher the epistemic risk. But the absolute minimum – corresponding to the strongest realist arguments – is achieved by focusing on material postulates themselves.

How content-driven are these arguments then? Do they go at all beyond scientific reasoning itself? Sure they do. The content-driven realist arguments are bona fide philosophical arguments. Scientists latch onto the correct material postulates by the methods of science, which may or may not make the material postulates transparent. If a scientist appeals to a theory *T* because it is the simplest and the most unifying, and hence the most explanatory perhaps, it is a task for the philosopher to make explicit how these contextual judgements reflect the particular material facts, given the scientific background knowledge of the domain in question. Only once material postulates have been made transparent can we compare them with the particular assumptions underwriting some commensurate inductions to the observable. Hence, although the specific content-driven arguments hang on case-dependent detail, its master plan can be described in general terms. This presents a new challenge for the philosophers of science, reforming the realism debate. The recurring question is: can we argue for realism about this, or that, in terms of local material postulates?

Notes

1. I'd like to thank Steven French and the editors of this volume for very helpful comments on an earlier version of this chapter.
2. This kind of selective skepticism is typically attributed to van Fraassen.
3. Admittedly there are many subtleties to Psillos's argument, regarding the rule-as-opposed-to-premise-circularity of the meta-level justification, for example, and the fact that Psillos of course allows for different degrees of confirmation: not all explanatory inferences are epistemologically on a par as far as their confirmatory strength goes. But the basic form of the justificatory argument is this, and it is form-driven.
4. Unlike Psillos, Lipton is not an 'IBE fundamentalist', claiming that all inductive inferences are best construed as abductive. Rather, for him the scheme of inference to the best explanation simply plays a significant role in understanding inductive reasoning.
5. Some of the *entity* realist arguments (e.g. Hacking, 1981) are also naturally interpreted as belonging to this category.
6. There's much to this paper besides the argumentative step that I will solely focus on. But this argumentative step, I take it, is the crux of Achinstein's argument against the selective skeptic.

7. Similar views have been championed by others before Norton. See e.g. Rescher (1980), Sober (1991).
8. I briefly mention this line of thought here because it usefully illustrates how engrained the more formal way of thinking about induction is. But I do not wish to get too embroiled in this debate (see Okasha, 2005, and references therein). For whatever its outcome is, my appropriation of Norton's descriptive thesis to the realism debate is equally valid.
9. Norton rejects the 'simple argument that such brute facts are always singular and that no collection of singular facts can license a universal', but doesn't explain on what grounds he does so.
10. Explaining the success of the scientific method by its truth-tracking ability is a philosophical explanation, albeit a naturalistic one.
11. For example, the fact that John handed his essay in on time is explanatory of John rather than Jane receiving the highest mark, given that Jane was penalized for not handing in her otherwise good essay on time. Handing her essay in on time is 'the corresponding event' absent in the causal past of Jane receiving her marks.

References

Achinstein, P. (2002) 'Is There a Valid Experimental Argument for Scientific Realism?', *Journal of Philosophy*, 99: 470–495.

Hacking, I. (1981) 'Do We See Through a Microscope?', *Pacific Philosophical Quarterly*, 62: 305–322.

Hacking, I. (1982) 'Experimentation and Scientific Realism', reprinted in M. Curd and J. Cover (eds), *Philosophy of Science: the Central Issues* (New York: W. W. Norton and Co., 1998), 1153–1168.

Harman, G. (1965) 'The Inference to the Best Explanation', *Philosophical Review*, 74: 88–95.

Josephson, J. R. (1998) 'Smart Inductive Generalisations are Abductions', in P. Flach and A. Kakas (eds), *Abduction and Induction: Essays on Their Relations and Integration* (Dordrecht: Kluwer).

Lipton, P. (2004) *Inference to the Best Explanation*, 2nd edn (London: Routledge).

Magnus, P. D. and C. Callender (2004) 'Realist Ennui and the Base Rate Fallacy', *Philosophy of Science*, 71: 320–338.

Miller, R. W. (1987) *Fact and Method: Explanation, Confirmation and Reality in the Natural and the Social Sciences* (Princeton: Princeton University Press).

Norton, J. (2000) 'How Do We Know About Electrons', in R. Nola and H. Sankey (eds), *After Popper, Kuhn and Feyerabend: Recent Issues in Theories of Scientific Method* (Dordrecht: Kluwer), 67–97.

Norton, J. (2003) 'A Material Theory of Induction', *Philosophy of Science*, 70: 647–670.

Norton, J. (2005) 'A Little Survey of Induction', in P. Achinstein (ed.), *Scientific Evidence: Philosophical and Historical Perspectives* (Baltimore: Johns Hopkins University Press), 9–34.

Okasha, S. (2005) 'Does Hume's Argument against Induction Rest on a Quantifier-shift Fallacy?', *Proceedings of the Aristotelian Society*, 105: 253–271.

Psillos, S. (1996) 'On van Fraassen's Critique of Abductive Reasoning', *Philosophical Quarterly*, 46: 31–47.

Psillos, S. (1999) *Scientific Realism: How Science Tracks the Truth* (London: Routledge).

Psillos, S. (2002) 'Simply the Best: a Case for Abduction', in A. Kakas and F. Sadri (eds), *Computational Logic: From Logic Programming Into the Future* (Berlin: Springer-Verlag), 605–625.

Reiner, R. and R. Pierson (1995) 'Hacking's Experimental Realism: an Untenable Middle Ground', *Philosophy of Science*, 62: 60–69.

Rescher, N. (1980) *Induction: an Essay on the Justification of Inductive Reasoning* (Oxford: Blackwell).

Resnik, D. B. (1994) 'Hacking's Experimental Realism', *Canadian Journal of Philosophy*, 24: 395–412.

Salmon, W. C. (1984) *Scientific Explanation and the Causal Structure of the World* (Princeton: Princeton University Press).

Sober, E. (1991) *Reconstructing the Past: Parsimony, Evidence, and Inference* (Cambridge, MA: MIT Press).

2

Optimism about the Pessimistic Induction

Sherrilyn Roush[1,2]

1. Why worry?

How confident does the history of science allow us to be about our current well-tested scientific theories, and why? The scientific realist thinks we are well within our rights to believe our best-tested theories, or some aspects of them, are approximately true.[3] Ambitious arguments have been made to this effect, such as that over historical time our scientific theories are converging to the truth, that the retention of concepts and claims is evidence for this, and that there can be no other serious explanation of the success of science than that its theories are approximately true. There is appeal in each of these ideas, but making such strong claims has tended to be hazardous, leaving us open to charges that many typical episodes in the history of science just do not fit the model (see, for example, Laudan, 1981). Arguing for a realist attitude via general claims – properties ascribed to sets of theories, trends we see in progressions of theories, and claimed links between general properties like success and truth that apply or fail to apply to any theory regardless of its content – is like arguing for or via a theory of science, which brings with it the obligation to defend *that* theory. I think a realist attitude towards particular scientific theories for which we have evidence can be maintained rationally without such a theory, even in the face of the pessimistic induction over the history of science.

The starting point at which questions arise as to what we have a right to believe about our theories is one where we have theories and evidence for them, and we are involved in the activity of apportioning our belief in each particular theory or hypothesis in accord with the strength of the particular evidence.[4] The devil's advocate sees our innocence and tries his best to sow seeds of doubt. If our starting point is as I say, though, the innocent believer in particular theories does not have to play the offense and propose sweeping views about science in general, but only to respond to the skeptic's challenges; the burden of initial argument is on the skeptic. The greatest strength of the realist attitude lies at this starting point, as Arthur Fine (1996) realized, and I will argue here that no one has given reason from the history of science to give it up. In particular, no pessimistic induction over the history of science has done what it needs to do in order to undermine our right to apportion our beliefs in particular theories to the evidence we have for them. To show this I am going to reconstruct what the pessimistic inductivist could possibly be saying when he tries to undermine our confidence, and show that there is one potentially threatening argument form he could use. However, this argument form requires much more than has been expected if it is to serve the pessimist, and pessimists have not actually ever offered what is needed.

For an example of how to start from the beginning, consider the realist of the sort I described, whom I will also call the 'optimist', who is typically confident that Quantum Theory has some very big things right. (As I will explain below, she does not have to know exactly which ones.) Her confidence in the claims of Quantum Theory is based on trust in the work of the scientific community, her understanding of the good evidence we have for this theory, and her own thought and knowledge about the physical world and logical and mathematical matters. There may be circumstances in which it would be helpful to her to have a further argument at her disposal that explains why all of this speaks to her having a right to her beliefs in Quantum Theory's claims. But the having of a right to a belief does not require her to have an argument available for defending just any challenge to the claim that she has a right to this belief. If it did then the justificational regress that follows would undermine our right to any of our beliefs.[5]

If this is where we start then the anti-realist will have to provide reason to believe that our innocent does *not* have a right to such beliefs. One way to do that would be to cast doubt on the optimist's evidence for Quantum Theory, meaning arguing about particular experiments and so on, and if those doubts were compelling then she would be obligated to respond to them. Of course, this has not been the strategy of the anti-realist, whose only other option, then, is to make a general argument for doubt that will apply to Quantum Theory *qua* theory, or *qua* theory of a particular sort. That is, the pessimist will have to find a fault or weakness that can be described without reference to the contents of specific theories. If he is to do this by an induction over history, then he must draw our realist into reflection about all (many, most, typical, etc.) theories and scientists of the past, and show why those reflections are both relevant and damning for us. One cannot do a legitimate induction from properties of swans to properties of bananas.

The pessimist's tools have to be at a general level and apply to past theories, but the confidence of the optimist that the pessimist needs to undermine is in our particular theories, e.g. Quantum Theory, so the pessimist needs to display connections both between general and particular, and between them and us. If instead of or in addition to history he is going to invoke general concepts, such as underdetermination of theory by evidence, then the anti-realist will have to explain why these things have an upshot for belief in Quantum Theory. I don't say these things cannot be done – sometimes a theory exhibits a simple instance of an epistemologically relevant concept or phenomenon, and surely there are some similarities between our predecessors and ourselves. I am making the point that this is what needs to be done in order to provide a challenge to optimism that we need to take seriously. In the pessimistic induction over the history of science the relevance of the failures of our predecessors is supposed to come by way of likeness. We have some property or properties in common with them that are relevant to whether one achieves true beliefs, and since they failed at the project of getting true theories, we have insufficient reason for confidence that we are succeeding.

I will argue that for the pessimist's historical evidence base of failure to be relevantly similar to us, his argument must be formulated as a meta-induction, an induction over our beliefs in scientific theories and evidence on the one hand, and the properties of those beliefs – truth, falsity, or fallibility, for example – where these take the position that swans and white respectively do in the generalization to 'All swans are white.' There is one way that similarity at the level of these meta-facts – they had beliefs ..., we have beliefs ... – could be the basis of the induction the pessimist desires, I will argue. However, simply because of the way inductions work, two more elements are needed: to know which facts about their beliefs are relevant to claims about our beliefs and why – the contents of the beliefs are different after all – and to know why even a meta-claim about ourselves, a claim about our having certain beliefs and dispositions to believe, should affect our confidence in our object-level claims, say of Quantum Theory. Although I will explain why the pessimist is on good grounds in the second claim, I will also show that finding a similarity that could support a possible induction from our predecessors to ourselves in the first place, requires recourse to properties so general that the induction is easily undermined by more particular and relevant differences.

Could the similarity between our predecessors and ourselves be at the level of evidence? I will look for a similarity at the first-order level – the content of evidence claims – to show that there is no way to avoid putting the induction at the meta-level. The content of our set of evidence claims is different from that of our predecessors, but the pessimist might hope to avoid that problem; our evidence *set* is different from theirs, but they overlap over some propositions. However, even particular evidence claims that we may share with our predecessors can be relevant to whether our theories are true only if their content is relevant to the content of our theories in the right way. Typically it is not.

The content of the evidence for or against most theories in history is not relevantly similar to the content of our particular evidence or theories. Obviously, whatever made the theory of bodily humors seem likely is not relevant to Quantum Theory in the right way. But the lack of appropriate relevance persists in more apparently plausible cases too. For example, whatever it was about the facts and the phlogiston theory that made scientists give up that theory is not evidence against our chemical theory or the parts of that picture that we might have retained. Our theory did not retain parts of that theory that we (and the pessimist) thought that evidence had falsified, so the evidence that falsified that theory does not falsify our chemical theory. The relevant evidence of our predecessors was supposed to be the evidence they had *for* their theories, one will say. But this fares even worse. Anything we did retain from their theory had not been falsified by any of our predecessors' evidence (or ours), so any positive evidence they had for their view is also positive evidence for that part of ours. The rest of the new theory we have about the subject matter would not be inconsistent with that evidence either, since we knew about that evidence when we formulated our theory. If the evidence for their theory told against our theory we would not be espousing our theory, since this is evidence that we knew about in choosing our own theories.[6]

These points depend on a comically idealized picture of the way transitions are made between theories over time. However, that fact does not help the pessimist. Consider the phenomenon of Kuhn loss, where a later theory fails to explain a phenomenon the earlier theory did. In such a case either our theory is consistent with that phenomenon (though shedding no light on it) or it makes predictions inconsistent with the phenomenon. If it is consistent, then that evidence shows not that our theory is false, but at most that it is incomplete. If our theory is inconsistent with that evidence, then there exists evidence against our theory. But the existence of evidence against Quantum Mechanics does not provide material for a pessimistic *induction*. There would be nothing essentially historical about the problem – at most history was used to discover a problem for a particular theory of ours, that there exists evidence contrary to it.

Or consider violations of what Feyerabend call the Consistency Condition, which says later theories must be consistent with the observational evidence for earlier theories and which I assumed above. As with Kuhn loss, such sets of evidence will not yield a pessimistic induction that casts doubt on Quantum Theory, since to find such a case is simply to find evidence contrary to our theory, not to find a phenomenon casting doubt on our theories that is essentially historical.

These points are sound when they concern particular theories and the contents of particular evidence, one might say, but what if Kuhn loss and violations of the Consistency Condition are recurrent phenomena in the history of science? In that case, the induction that may be possible is a *meta*-induction, since it concerns similarity of our predecessors to us in the believing of theories for which they and we had (have) good evidence, and in the existence of evidence against their theories that they were, and therefore we probably are, unaware of or stubborn about. A theory can be mostly or approximately true, however, while not being consistent with all evidence – indeed it can be exactly true despite some contrary evidence, provided only that we got an auxiliary a little bit wrong. And we can be legitimately highly confident in a theory while also being highly confident that something is wrong somewhere without knowing where, and thus legitimately confident that there is some contrary evidence somewhere too. I will explain these points when I come to the preface paradox below. If the pessimist could show that the phenomena of Kuhn loss and violations of consistency are large and rampant throughout the history of science, then he would be on his way to the meta-induction described. However, the facts about method that I argue below undermine the last possible pessimistic induction would also undermine that one.

These points and common sense tell us that the relevant similarity between us and our predecessors are the general facts that our predecessors *had* evidence that *supported* their theories. Since their evidence was inadequate to their presumed goal of having true theories, the similarly supportive evidence we have is also likely inadequate to the same goal for our theories. The similarity for a potential induction must be between how their evidence and beliefs related to their theories' truth values and how our evidence and beliefs relate to our own theories' truth values. Their beliefs were somehow justified by their evidence and false, and our beliefs are also justified by our evidence and despite this also probably false.

The problem with this approach is that in escaping the differences in particular content of theories and evidence from the past to our own by going to the general comparison, the pessimist will be making an induction that falls victim to those same particular differences in a different way, because of the way induction works. It is an elementary and familiar fact about induction, that even if all F so far are G, the inference to the claim that the remaining cases with F will be G is rendered illegitimate if the remaining cases are known to have a further property which we know is relevant to whether G arises. Reichenbach called the move where we take into account that further evidence 'cross induction' (Reichenbach, 1949: 430). Thus, the epistemological mistake of the man who jumps off the top of a building and at the 40th floor says 'So far, so good!' is not in making the induction in the first place, nor in being wrong in his conclusion, but in failing to use his knowledge of terrestrial gravity to undermine the induction with a cross induction.

There is straightforward material for a cross induction against the latest argument I have described on behalf of the pessimist. If the pessimist cites for the base property of inductive comparison the fact that our predecessors, like us, had (good) evidence for their beliefs, then two facts combine to counter the inference from this similarity and the falsity of their theories to the conclusion that ours are probably false too. One is that the content of our evidence propositions and evidence sets is different from the content of theirs; at the very least there is *more* of it. We have done more and different experiments, for example. Secondly, the content and quantity of evidence is relevant to whether the belief one bases on it is likely to be true. This is so, at least, unless no amount of evidence short of infallibility is sufficient to justify any degree of belief in a theory, and unless more or different evidence does not make a difference to whether your belief based on it is likely to be true. But the problem those possibilities pose is the general problem of induction, and if pointing to that was the pessimist's argument he need not have reported it anew or enlisted history to make it. Moreover, it is an argument that would undermine his own induction. Since our evidence set is relevantly different from our predecessors', it remains to be shown why their failure is a reason to believe in ours. The optimist is in a strong position here, for she has no burden to argue that our evidence is *better* than that of our predecessors. It is the pessimist's burden to say why it is not.

The first meta-induction I have reconstructed for the pessimist is undermined by a simple cross induction. However, there is another meta-induction available. In this argument we would take the base property again to be believing and having (good) evidence, but the inferred property will be not being wrong but being unreliable. If our predecessors' theories were usually false, as the pessimist supposes, then our predecessors were *unreliable*. If we, like them, believe on the basis of what looks to be good evidence, there would seem to be excellent reason to think we are unreliable too. The unreliability of belief in theories is different from theories being false because it is different from a belief being false. A belief can be unreliably formed and yet happen to be true. Thus, because the property of unreliability is distinct from that of falsehood, this holds out hope that there is a good pessimistic argument after all. What would follow about our rational confidence in particular theories if the pessimist succeeded in this way in giving reason to believe we are unreliable is again a separate question.

In what follows I will address the argument just described, and the proper relationship between beliefs in our reliability level and beliefs in our theories. For this purpose we need a more precise understanding of the properties of reliability and fallibility, and the proper relation of beliefs about whether we have these properties and rational confidence in our theories' claims about the world.

2. Fallibility

There is no disagreement today that our science is fallible, that is, that we *could* be wrong, that our evidence is not strong enough to imply the truth of our theories. The issue is not whether we are fallible, but whether given that we are we can nevertheless have a right to confidence that our theories are true. The pessimist just imagined tries to show us we don't by showing that we are not just fallible but *unreliable*. But one might think he does not need to go that far, for is there not just something wrong with admitting you are fallible on the one hand, and persisting in high confidence in your theories on the other?

If this is our reason for pessimism, and if it works, then we do not need an induction over the history of science. However, showing what is wrong with this idea in its various forms will yield an understanding of fallibility that will in turn allow us finally to evaluate the pessimist's induction as most recently described.

To see what looks wrong about acknowledging fallibility and remaining confident, suppose that admitting our fallibility means being confident that at least one of our claims or theories, p_1, p_2, p_3, ..., $p_{10,000}$, is wrong:

* not-p_1 or not-p_2 or not-p_3 or ... or not-$p_{10,000}$

How then can we simultaneously be confident in each of those claims

** p_1, p_2, p_3, ..., $p_{10,000}$

when their conjunction

*** p_1 and p_2 and p_3 and ... and $p_{10,000}$

directly contradicts the disjunction * above? On the other hand, one can imagine what looks like a quite sensible way of having such an array of beliefs by imagining a person who confidently offers the claims of her book, yet also in the preface admits that a few of them are probably wrong. The difficulty here has come to be called the Preface Paradox. Philip Kitcher (2001a, 2001b) noted some years ago the resemblance of this paradox to imagining our beliefs in scientific theories as justified despite our acknowledgement of possible error, and sided with intuition. Of course the person writing the preface is saying something sensible, whatever precisely we eventually figure out it must be. She is not contradicting herself and we are not either when we do not take our fallibility to be a reason to give up our confidence in our theories.

This is right in my view, but it would also be good to know precisely what she is saying, so that we can put to final rest the sense of suspicion about our confidence in scientific theories that this paradox nurtures. Over the years a variety of proposals have been offered for resolving the Preface Paradox, but they carry high price tags. Some have involved denying that conjoining conjuncts preserves justification (denying closure of justification under conjunction). Others propose that the lesson of the Preface Paradox is that it is possible to rationally believe two contradictory claims – the conjunction and the disjunction – as long as we avoid conjoining them and thereby believing a contradiction. Fortunately I do not think we need to go this far, for if we take it that rational degrees of belief are required to satisfy the probability axioms, thus,

S rationally believes q to degree x iff, $P_S(q) = x$,

where P_S is a probability function, then we will be able to see that there is no paradox at all. The facts I am about to discuss are familiar to probabilists, possibly so familiar that it explains why I have not been able to find this solution written down. But it is worth putting them down in a way that highlights their salience to our questions here.

I take it that the attitudes of the preface writer that we need to explain involve being highly confident in each of the claims of her book, and also highly confident that at least one of them is wrong. But it is simple to show that if one's degrees of belief conform to the probability axioms and they are not extreme – zero or one – then having a preface-writer set of attitudes is not just permissible, but obligatory. Suppose I am highly confident in each of my book's claims:

** $p_1, p_2, p_3, \ldots, p_{10,000}$ $P_S(p_n) = .95$, for each n

Then I am rationally required also to have a very, very high degree of belief that at least one of them is false, i.e.:

* not-p_1 or not-p_2 or not-p_3 or ... or not-$p_{10,000}$ $P_S(*) = .99$ repeating

This is because the probability of a conjunction, here

*** p_1 and p_2 and p_3 and ... and $p_{10,000}$

is the *product* of the probabilities of the conjuncts.[7] Since the confidence I have in each conjunct is not perfect but .95, the required confidence in the conjunction is very low indeed. I would only need to have 59 such conjuncts in order for the required confidence in the conjunction to drop to 5 percent. 95 percent raised all the way to the 10,000th power, as in this case, makes the probability of the conjunction practically nil. It immediately follows that the probability of the negation of this conjunction, that is, of the disjunction of the negations of the conjuncts, must be very, very high. This is just how the degrees of belief of our preface writer are arranged, with her very confident that at least one of her claims is false.

One might object that on this picture the rational credence in the conjunction of claims of the book is still low, whereas the preface writer should surely be confident of that conjunction. But as long as we assume that her degrees of belief are not extreme, the probability axioms require her to have very low confidence in the conjunction despite her very high confidence in each conjunct, so this objection amounts to insisting that the preface writer should be probabilistically incoherent. The strongest of the author's reasons to believe each of the p's and their connections will anyway tend to be local, and to ask for her confidence in the conjunction we would have to ask whether she thinks *all* of the claims true. Given the obvious contradiction between this and the claim that at least one of the claims of the book is false, surely she would demur.

What if the preface writer's degrees of belief concerning these matters are extreme, zero or one? If her confidence in each of the conjuncts is full then probabilistic coherence requires her confidence in the conjunction also to be perfect ($1^n = 1$), maximal, and thereby also requires her to have zero confidence in the disjunction of negations of the conjuncts.[8] This may look like a problem, since for such a person to be probabilistically coherent, she must not have the slightest degree of belief that one of her beliefs is false; she must not acknowledge her fallibility, so she must not write that preface if she is to remain rational. But while this is true, it is not paradoxical, for the person described in this case is not plausibly someone who *would* write that preface in the first place. The person we are imagining has perfect confidence in each and every claim of her book. This means she does not leave open any possibility of being wrong, the formal expression of which is the fact that the maximal degree of belief (one) makes impossible any revision of that belief on the basis of change in other beliefs. The paradox of the preface does not arise merely from the fact that p_1, p_2, p_3, ..., $p_{10,000}$ are together incompatible with not-p_1 or not-p_2 or not-p_3 or ... or not-$p_{10,000}$. There also must be reason to suppose that those propositions (or other appropriately contradictory relatives of them) are ones that the writer of the imagined, apparently reasonable, preface is confident of in an incoherent way.

One might object that how confident one is allowed to be in the disjunction of negations depends, on this picture, on the number of conjuncts asserted, and must be small if there are few conjuncts. The confidence of our preface writer, on the other hand, that she has made at least one error is surely high. The allowed confidence in the disjunction does indeed depend on the number of conjuncts on this picture. However, the preface writer's level of confidence that she has made at least one error also intuitively depends on how many claims she has made. If you had confidently made only three claims, how confident would you be that at least one of them is wrong? Surely not a great deal more than your hesitancy in each of them, in conformity with what the probability axioms require. In such a case it would be strange for the preface writer to express high confidence of error. Conformably, I have never seen an abstract or first paragraph of an academic paper announce the high likelihood that something in the paper was erroneous. The reason that the Preface Paradox is most pressing in the case of a book isn't just that books have prefaces, but also that books are long.

This straightforward resolution of the Preface Paradox depends on avoiding a threshold conception of justified belief. That is, we do not ask whether or not a person is justified in (fully) believing p depending on how high the probability of p was taken to be, or at what level of evidence quality a subject is justified in having full belief. Rather, we ask whether or not a person is justified in having a certain degree of belief in p, and this depends on whether that degree is probabilistically coherent with the rest of her degrees of belief in evidence – and other propositions. Thus, a person is justified in having *full* belief in p if and only if her evidence and other beliefs require her belief in p to be one (see note 5). The reasonableness of this solution is seen in how well it incorporates the phenomena of fallibility, which was what the threshold conception was trying to do in taking a non-maximal amount of evidence to justify a maximal degree of belief.

Just as it is rational to write the preface while remaining confident in each of the claims of your book, it is rational to acknowledge that some of the claims of your theories are probably wrong – you know not which – and yet to remain confident in each of those claims. A conception of rationality as requiring probabilistic coherence wouldn't have it any other way. In particular, we are well within our rights to be both confident in Quantum Theory and also confident that something of it probably mistaken, without needing to know where the problem is. Thus, confidence in our theories can coexist with confidence in our having made some mistake. This is another expression of the fact that the pessimist cannot win at the first order but must deal with claims we make *about* our beliefs and theories.

There is a richer way of representing the claim of fallibility than the confidence in the disjunction of negations of my beliefs, and it will give our pessimistic inducer another chance. It is of course not impossible for a maximally confident person to write the self-doubting preface; only, as we said, implausible. If we express fallibility in a more adequate way than we have so far, we will see that it is not even first-order probabilistically incoherent for the maximally confident person to write that preface. There is a broader irrationality involved in it though, that corresponds to our expectation that a person would not feel inclined to do that. The formulation of fallibility above implicitly took my expression of my fallibility about the topic of my book as itself in direct competition with my confidence in each of the propositions in question, equating my recognition of my fallibility with a lack of perfect confidence that all of those propositions were true. This is one way to understand the problem, and it generates the most direct version of the Preface Paradox that we just dealt with. However, what I am asserting when I say with high (not maximal) confidence:

$$\text{not-}p_1 \text{ or not-}p_2 \text{ or not-}p_3 \text{ or} \ldots \text{or not-}p_{10,000}$$

is not giving me or my audience any more information over my lack of perfect confidence that:

$$p_1 \text{ and } p_2 \text{ and } p_3 \text{ and} \ldots \text{and } p_{10,000}$$

(except the fact that with regard to these degrees of belief I am coherent and responsive to the relation between the contents of these propositions). My high, or at least non-zero, confidence in the disjunction of the negations of the conjuncts is not a statement of my fallibility. It is simply an expression of my less than perfect confidence about particular matters.[9]

When someone discovers that he is sometimes wrong about such things, what he discovers is unfortunate, his discovering it is fortunate, but neither is just the same thing as his lacking confidence in the particular propositions. Rather, it says he has a general property, a tendency among his beliefs; it is a statement *about* his beliefs, a statement which would include among its terms not only 'p' but also 'B(p)', meaning the subject believes p; as such, a statement of my own fallibility is second-order, expressing a belief about my beliefs. We feel that the maximally confident person should do something in response to this discovery about his beliefs, but it is not impossible as a matter of fact for him to fail to. It would not even make him incoherent at the first-order if he failed to respond with a reduction in his confidence in p (not without further assumptions). This way of thinking about fallibility thus avoids the first-order Preface Paradox too. However, intuitively, believing ourselves fallible surely places some kind of rational demands on us in the management of our first-order beliefs.

What exactly those demands of rationality are is a general, non-trivial, issue, but this is a question the pessimistic inducer needs an answer to. For even if he succeeds in convincing us that the unreliability of our predecessors is a reason to believe in our own unreliability, we now see that we must ask *why* this should obligate us to reduce our confidence in Quantum Theory. Here I will do the pessimist a favor by filling in this blank in his argument by defining fallibility as a second-order property, and explaining why and how it places rationality constraints on first-order beliefs.

If I am asserting an explicit claim of my fallibility in the preface, I am not withdrawing confidence in my beliefs but making an observation about the probability that creatures like me, with dispositions like this, get it right about matters like this. The claim that q and the claim that I or those like me have a general tendency to imperfect beliefs on q-like matters, are distinct kinds of assertion.

It seems clear to me that the second-order property that is fallibility is an inverse of reliability, which is also a second-order property. We can say that a person's belief-forming process is *x percent reliable* (with respect to q) when:

$$PR(q/B(q)) = x$$[10]

That is, the probability that q is true, given that the subject has full belief in q is x, where for simplicity we are taking 'B(p)' to mean perfect confidence.[11] For the moment we may think of this probability as a rate of success, representing how *often* q is true when the subject believes it; an interpretation suitable in this context must in any case be objective, hence 'PR' instead of 'P'. Obviously, the probability that a person is right about q when she has perfect confidence in it is not necessarily 1; confidence does not imply reliability.

From this definition fallibility can then be expressed as the size of the gap between her reliability and perfect reliability. Perfect reliability means the rate, PR(q/B(q)), equals 1. Always, when she is sure of it, it is true. Thus, she is fallible (in her full beliefs) to degree y when:

$$1 - PR(q/B(q)) = y$$

To acknowledge one's fallibility is to acknowledge that $y > 0$, and that $PR(q/B(q)) = 1 - y$. That is, it requires a degree of belief in a statement about the objective conditional probability PR(q/B(q)). However, without further assumptions one's belief about this conditional probability does not constrain one's belief about q in any way.

When we say we might be wrong somewhere, we could mean this second-order claim. It does not immediately conflict with our first-order beliefs, which is good news for avoiding a new second-order Preface Paradox. The bad news that does suggest a new Preface Paradox is that it would be broadly irrational not to respond to the discovery of fallibility by doing something about our confidence in q. We will resolve this below in a way that reassures us in both of these directions. To make the connection between the two orders we would need to make an assumption about how a degree of belief about a property of our degree of belief in q should relate to degrees of belief in q. Since these two beliefs are distinct states of affairs at different orders, probabilistic coherence alone – conforming one's degrees of belief at a given order to the probability calculus – does not by itself give a constraint concerning that connection. For those in the know, note that this is explicitly acknowledged in the fact that $P(q/P(q) = x)$ is taken to be an independent constraint worthy of discussion and disagreement, and requiring argument.

3. Non sequitur

We now see how to express admission of fallibility in a way that is *consistent* with first-order assertion of particular claims, but is it not clear that the admission still undermines our justification for the first-order belief by an obvious and immediate inference? Wouldn't our serious fallibility as scientists, if shown, automatically give reason for an equally serious reduction of our confidence in Quantum Theory?

We should not conclude this hastily. Consider one consequence if we did think that the admission of fallibility immediately undermines our justifications for first-order beliefs, even domain-specifically. Most of us will have been frustrated at least once by a typical Creationist (or Intelligent-Designist) line of rhetoric one hears, which points out that the theory of evolution has not been *proved*. Therefore, it is concluded, opposing views (especially one in particular) have a right to be taught in public schools. We should have an 'open mind', and not be so confident in the views of the scientific mainstream that we present them as true. It certainly looks like the Creationist is appealing to the fallibility of science in this argument, and taking it to require no more than 50 percent confidence in evolution.

One obvious reply to this is that he is mistaking a small quantity of fallibility for a problem so large that it puts all views and frameworks on an equal footing of plausibility. 'Proof' is for mathematics and logic, not for empirical research where fallibility is just the way it is. The possibility of error does not justify an explosion to the night in which all cows are black. Our theories are fantastically better studied and verified than the opposing theories vying for a platform in public schools. But how can we justify this claim about the irrationality of the move to equal plausibility? And if the level of fallibility is significant, then what? The resemblance of the form of the pessimistic induction over the history of science to this Creationist rhetoric is somewhat alarming. In charity we must suppose there is a difference, but it would be nice to know what it is. We will see below that when we understand what has potential in the pessimist's argument – by understanding the rational place of fallibility claims – we will also see what is fallacious about the Creationist argument.

The pessimist has potentially more to offer than the Creationist, if he can convince us that our predecessors were often or always wrong, meaning they had a low, or very low, reliability, and also that we are similar. But how should we count? What if there are more theories in the last 100 years than in the previous 5,000, and whereas all of the previous ones were false (or at least abandoned) a high percentage of those in the last 100 years are still not falsified (and are retained)? All of these theorists are our predecessors and the overall numbers wouldn't yet be shown to be bad on this quite plausible way of counting. It is not clear how anyone will decide non-arbitrarily how to count the historical cases, but this is a problem for the pessimist now since we have put the ball in his court again by describing one plausible way of counting that makes the fallibility of our predecessors low or unknown.

In charity to the pessimist I will suppose we can find a reason to think the supposedly large impairment of our predecessors is shared by us, in order to reconstruct his best possible argument. The current problem is why even if we successfully attribute a surprising, significant, general level of unreliability to ourselves, this should issue in a withdrawal of our confidence in specific first-order matters, such as particular claims of the Special Theory of Relativity. Suppose that q = the speed of light is not different in different frames of reference, where frame of reference is defined by velocity. The claims that are probabilistically relevant to that claim are the results of the Michelson-Morley experiment, the claim that the interferometer arms were equal in length, and so on. The first is relevant because if the experimental apparatus works as it should then the fact that no interference fringes show up in the experiment raises the probability that the speed of light is constant over reference frames. The second is relevant because length equality in the interferometer arms increases the probability that the apparatus works as it should, i.e., that fringes you may see are due to variation in the speed of light. Length equality is relevant to the hypothesis because it raises the probability of that hypothesis given the experimental outcome. Our being fallible is definitely not relevant to the Special Theory of Relativity in the way that typical evidence like this is.[12]

As pointed out above, it is a straightforward fact that the general level of unreliability we might attribute to ourselves,

$$PR(q/B(q)) = x,$$

is not probabilistically relevant to

$$q$$

and thus not to how much confidence we should have in q, without further assumptions. PR(q/B(q)) *is* related to PR(B(q)) via:

$$PR(q/B(q)) = PR(q\ .\ B(q))/PR(B(q))$$

But the question is what confidence we should have in q, given our credence for PR(q/B(q)) = x, not what confidence we should have in B(q) in that circumstance. How confident should I be in q if I believe that the probability that q is true given that I believe it is x?

4. Descent

I will call this the problem of 'descent' because we have to say how and why a second-order discovery about our tendency to believe should affect our confidence in specific first-order claims. Surely a discovery that, say, we fall into a class of people who have a 20 percent error rate on theoretical claims doesn't mean we need to drop our confidence in the relativity of simultaneity to 50–50, meaning that all bets are off. But it is reasonable to think it means *something*.

A notion that would give the pessimist's intentions their due is that of calibration: we are calibrated when we match our confidence to our accuracy (reliability). We can see an example of the concept of calibration in the way that an eyewitness's confidence on the witness stand relates, or should relate, to his other properties. We onlookers may have no information about the person's track record in making face recognition judgements, but there is a track record and it appears to be epistemically good, in some sense, if the witness is no more confident in his judgement now than the percentage of times he's gotten such judgements right in the past. His track record is a clue to how reliable he is. If he's been right 99 percent of the time it would be epistemically good if he were very highly confident. (Otherwise in trusting him only to his level of confidence we are deprived of the information he likely has.) At the other end of the spectrum, if he were right 50 percent of the time in past such cases, we hope that he would say 'I don't know.' He may have no inkling of his bad track record – he may have acquired amnesia since his last such judgement – and if we don't know about the amnesia or the track record we are out of luck. It is clear what would be most epistemically sound and helpful in these situations, from imagining it present and imagining it absent.

There is psychological evidence that jurors use confidence of a witness as an indication of accuracy by default, effectively assuming calibration without any information pertinent to the property. That is, on first encounter with a witness, a juror tends to believe what a confident witness says in direct proportion to the witness's confidence. Since confidence implies nothing without calibration, this appears to be an epistemic disaster for the system of trial by jury, and was greeted as such on its discovery. This illustrates both that calibration is a non-trivial property, i.e. that confidence and accuracy are distinct properties that do not necessarily line up in naturally occurring beliefs, and that it matters to us; it would be good for us if a witness's testimony were calibrated before we form any particular degree of confidence in what he says, even if, as a matter of fact, we tend to be poor judges of calibration. Incidentally, and in keeping with the idea that we do demand calibration, it has recently been discovered that (mock) jurors do make use of any further evidence of error by that witness that comes available, to reduce their judgement of his calibration and thereby of his credibility when he is confident (Tenney et al., 2007). The jurors' default assumption of calibration is defeasible. We not only care about this property of calibration but implicitly reason with it continuously in dealing with new evidence and evidence about those who deliver it.

So, suppose calibration – having your confidence match your accuracy – is a good thing, and that by 'accuracy' we mean reliability, probability of being right when you believe, the property defined above that we can get information about by looking at track record. This would imply that our confidence in our particular theories should match our reliability in making such judgements, and that if we aspire to that then our confidence should match what we *believe* our reliability is.[13] We have hereby uncovered the basis of an intuition that makes us feel a pull from the pessimistic induction: if the pessimist can show us that the (supposed) low reliability of our predecessors is good evidence that *we* have a low reliability, then we should reduce our confidence in our particular theories to the level of reliability we have now come to believe we have. If we believe our reliability level as scientists is 50 percent, then we should be no more confident than not that our theories are true. Thus, we can see how the pessimist can finish his argument on us, if he manages to convince us we are unreliable. But this is the only good news for the pessimist. In what follows I will discuss how the historical evidence the pessimist gives us *could* be a reason to believe we are unreliable, but argue that it is not.

Note that this way of viewing the relevance of the second-order property of fallibility to what our first-order confidence in particular theories should be does not license the Creationist move. This is because the rule I have described does not imply that fallibility implies equal plausibility of all theories. The fallibility will, speaking in idealization, be a given number, the reliability being 1 minus that number. Suppose, for illustration, that the fallibility level of contemporary science is 20 percent; then the reliability level is 80 percent. The rule says match your confidence to your reliability, so in this case our confidence should be 80 percent. Discoveries about your reliability level in q-like matters may obligate you to push your confidence in q up or down by a small or large amount, but they do not automatically mandate, or even license, a leveling in which every view is as good as every other. This view also shows why the amount of fallibility matters – our allowable confidence varies with it in a way that conforms to the fact that .02 fallibility is quite a bit better than .3 and almost as good as 0 (perfect). In so doing it also resolves the second-order Preface Paradox, by showing that you can rationally both believe you are fallible and maintain confidence in your particular theory, in the right proportions, of course.

5. What have they to do with us?

Suppose our predecessors had low reliability, meaning that they had a pretty high rate of believing things that were false. If our reliability level is the same as theirs then, because we should be calibrated, we should have low confidence in our theories. The question is whether our reliability level is the same or similar, and whether theirs being low is a *reason* to think ours is so. If I have been right above, then this is the best way of seeing the issue at stake in the pessimistic induction. The relevant question is not whether their theories were false and whether that should matter to us – as we saw above the pessimist can't get where he wants to go that way. He will instead argue that because they justifiedly believed and were unreliable our justified beliefs are unreliable too. Note that the more particular fact that we have *different* evidence for our theories than our predecessors had for theirs, which undermined the pessimist's induction above, will not do the job of saving our optimist here. The content of evidence claims cannot be depended on to be relevant to whether our beliefs are reliable since we could easily churn out different evidence in each round of research, but maintain the same miserable level of unreliability, for example because we use the same sorry ways of gathering evidence. However, there are particular second-order facts that will undermine the new induction, which involve differences in the use of methods over time.

Let us suppose that we are licensed in concluding that our predecessors are unreliable by looking at their track record. If we are to block the inference to the conclusion that we are too, it would seem that we need to know something relevant to our track record. And one might wonder how we could investigate our own reliability level without already knowing whether our theories are true. To see our reliability by rate of true theories we would need to do a count among the theories we believe, and see how many are true. But theories that we know are false we no longer believe, and asserting the truth of any or all of the ones we believe would seem to be begging the question here. If there exists any discernible track record for us, must it not be in the broad sense of 'us', the set of human beings doing science, in which case, our having granted the premise that our predecessors were unreliable would mean the pessimist wins after all.

What this misses is that the thing we need an estimate of is not track record *per se* but reliability. Tallying a track record is one way of estimating reliability but it is not the only way. For example, we can reasonably judge a machine to be reliable at doing its job through knowledge of its mechanism, or we can justifiably judge it as more reliable than another machine via our knowledge of the two mechanisms and of how good such mechanisms could possibly be at delivering the result we want. Another kind of simple example to make the point: we know the valences of hydrogen and carbon. Say we know the valence needed at the active site of any enzyme that is going to catalyze a chemical reaction we want to know how to produce. Suppose carbon has that valence and hydrogen does not. A given molecule with a carbon sidechain may not work, but we can see that its chance of doing so is fantastically higher than that of a molecule with only hydrogen sidechains.

We see an analogous thing in comparing our science to that of our predecessors, if we think of methods as mechanisms for leading us to believe the truth and avoid falsehood. Over historical time, our reliability is potentially and probably different from that of our predecessors because we use different methods. Here I use the term 'method' broadly, so that it includes, for example, techniques that do not take the form of rules, and also includes techniques that are specific to a given subject matter. The greater the historical time between us and them, the greater the probable difference in methods between us and our predecessors. If so, then this will be the material for a cross induction of the latest pessimistic induction if I can show that methods, in particular those that have actually developed over historical time, make a difference to reliability.

It is emphatically not necessary for the optimist to show that our methods are better – if we show that our methods are different and that methods are relevant to reliability, then it will become the burden of the pessimist to show that our methods are not better, and that thereby we are subject to the induction from our predecessors' unreliability. However, we can, I think, argue for the general superiority of later methods to earlier methods in many cases, and this will be the easiest way to show the weaker claim we need, that method makes a difference to reliability.

There is something to the stereotype that the Ancients preferred speculation to observation, and that speculation combined with observation is better than speculation alone. There is something right about the idea that Bacon's interventional experimental method, and his rules for safeguarding against psychological prejudices, get us more information than mere casual observation, at least, if any method gets us anything at all. There is something to the idea that following the refinements of experimental methods like those described by John Stuart Mill and William Jevons gives us more accurate and safer beliefs. The twentieth century brought a massive number of discoveries and refinements in statistical methods, and the new century has brought further ones already. Prima facie it looks pretty good for a claim that in reliability we are not as bad off as our predecessors. A more detailed look bears this out.

Consider some examples of methods, e.g. simple induction and cross induction. We may be able to show that induction brings us truths in the long run if any method (e.g. counterinduction) does (Reichenbach, 1949; Salmon, 1967). We may be able to show that the straight rule of induction is faster than any other method of ampliative inference (Juhl, 1994). But it does not seem that we can show flatly that induction is likely, or even *more* likely than other methods, to bring us true beliefs. It may be that none are getting onto the world. However, as we have noted, if the pessimist is going to hang his case on the fact that every scientific enterprise faces the problem of induction generally, he could have left off his induction over the history of science. We can show that problem through one toy case. Moreover, the pessimist usually does want to admit that science gets us some more restricted kind of knowledge than that of theories, since otherwise he has a hard time explaining the worth of what scientists are doing.

Let us assume that the world is susceptible to inductive procedures. I will argue that some methods are more reliable than others – where reliability is as defined above, the probability that q is true given that one has come to believe it by that method. In particular, some of the methods we use are more reliable than any of the methods our predecessors used. To show that a method is more reliable than another method, it is sufficient to show that there are potential errors that the first method is more likely to catch than the second method is, and that there are not errors that the second method is more likely to catch than the first method is. If this is to be sufficient for defending ourselves against the foibles of our predecessors, we must assume that the application of the first method is just as competent as the application of the second. However, the pessimist hasn't shown that our application of method to the world is worse than that of our predecessors, so we are free to make this assumption.

First let us consider methods pairwise on a question about a connection between two properties, F and G. We could construe the question as whether the one causes the other, or simply whether the correlation will continue as we look at more cases with F. We will apply both methods to a finite data set in which every case with property F also has property G. Now consider induction by generalization. This simple rule says that if all the cases you have seen having property F have property G too, then infer the generalization All F-cases have G. This rule is subject to the simple kind of error of failing to make an available cross inductive move such as, for example, in the case of the flying man above, using his knowledge of gravity. It is hard to imagine a human being so uninformed about the world that he doesn't know independently of his jump that the sequence of correlations he is seeing must shortly come to an end. But to get himself to a warranted belief about whether or not he will go splat, it is not enough to know this; he must also use the information. The method of induction by generalization did not tell him to do that. Adding the method of cross induction would. And if you use any information you have that is the basis for another induction that cuts across the trend you see and shows that it cannot, or cannot be expected to continue, there are many cases in which you will more likely avoid a false belief than your counterpart who does not supplement induction by generalization with cross induction.

The error that cross induction guards against is precisely one that would reduce our reliability in the sense defined above, for it would be a case of the thing you believe (that your fall is not dangerous) being false. The point here generalizes since any induction is subject to potential cross inductions due to the non-monotonicity of ampliative evidential support; unlike a deductive inference, an inductive inference is erodable by addition of premises. It is true that any cross induction is also susceptible to undermining by further evidence, but that is a new error whose discovery will also require employment of the method of cross induction. It does not immediately appear that there can be potential errors of the method of induction plus cross induction that are caught by the method of induction alone, since the former method incorporates the latter, and the interaction between the two parts of the former method does not seem to create new error possibilities.

One might object that any cross induction adds one more piece of evidence to the premises, which will make the premises as a group have strictly more potential error than the inductive inference that evidence was supplementing.[14] And surely it cannot be proven in general that that added error potential is less than the error the cross induction allows us to avoid. This is true, but the comparison to be made is between a method in which the next use of evidence is always for making a straight induction or the next use of evidence could issue in the making of a cross induction instead. On both sides of the comparison you are adding a piece of evidence. The evidence added in the two types of case could possibly have different potential error themselves, but we assumed above that we are no worse at *applying* our methods than our predecessors were at applying theirs so we may plausibly assume a symmetry between what they and we would achieve in error potential of our evidence.[15] The pessimist could try to argue that we are worse at applying our methods, but that wasn't a position I have heard a defense of. Thus, induction with cross induction is more reliable than induction alone. Moreover, the idea that this is so only by our own lights, a favorite phrase of pessimists, has a hollow ring.

A contemporary Bayesian approach to method incorporates the cross-induction rule in the following way. An idealized agent's beliefs are represented as a total function from all the propositions of a language to numbers between zero and one that represent strength of belief, that is, as a total probability function. This means that this ideal agent has degrees of belief about all matters, and most importantly here, that all the beliefs that any agent actually has are expected to be taken into account in any application of any rule that falls out of the probability axioms. The probability calculus thus incorporates a general, rigorous version of the cross-induction rule, and it brings more. For example, abiding by it allows us to expose equivalence relationships between different formulations of the same and related information, thus helping to identify cross inductions we might otherwise have missed.

The quantitative aspect of the Bayesian approach also yields rules for weighing how far a cross induction undermines an induction. In the case of the flying man the undermining is complete. Not so in many other cases. In coming to a conclusion in such cases, we are better off if we can tell whether the first induction or the cross induction is *more* powerful, and this approach provides techniques for doing that. This is only to mention some specific features of Bayesian method that are relevant to the execution of cross inductions. Advantages of further features and methods would take several books to describe, but we don't need to consider those to make the point that some methods are more reliable than others, and in particular that some of our methods are better than some, and in some cases any, of those of our predecessors.

Another problem with simple induction noticed probably first by Francis Bacon, and dealt with increasingly rigorously and expansively by John Stuart Mill, Charles Sanders Peirce, Neyman and Pearson, Sir Ronald Fisher, and many others who have followed, is that in the investigation of how potent an observed association between F and G is, one must deal with the possibility that the effect G was brought about by something other than F that one was not aware of. This is now addressed through an enormous and growing list of techniques: methods for insuring a representative sample of F's, the setup of a control group that doesn't have F in order to see whether F was really what made the difference to G's presence, the randomization of the control group to attempt to insure that variables other than F are not causing any correlation seen. And the foregoing are primitive descriptions of old techniques for squeezing information out of experiment and observation, that have been enriched and supplemented in the last fifty years by entire fields of researchers, in departments of statistics for example, and even among philosophers and computer scientists in the case, for example, of causal net programming. To bring the relevance of method to reliability full circle, scientists have increasingly acquired more, and more sophisticated, methods for evaluating the reliability of their applications of method.

In many of these cases we do not need to appeal to track record to determine whether one method is more reliable than another at achieving a particular aspect of the epistemological goal. Even when the statistical community has not settled a matter – e.g. between Bayesians and classical statisticians over whether a randomization procedure provides a better control group than mere matching – it is not imagined to be helpful to try to settle the dispute about method by pointing to a track record of scientists getting the right theory when they use it. Thus we do not need to already know whether or not our scientists are finding true theories in order to evaluate the quality of the methods they are using compared to the methods of their predecessors. The bottom line for our argument here is that these sophisticated methods are used today and were not used by our predecessors in earlier centuries; method is relevant to reliability – as we have shown by showing that some later ones are more reliable than some or all earlier ones, and this is material for a cross induction against the inference from our predecessors' unreliability to ours.

6. More is all we need

More, and more specific, investigation along these lines of comparing methods over the history of science would be valuable for judging where we stand, and this, I suggest, is what the discussion of realism and anti-realism might constructively give way to, or at least give a place for. At a gross level we know that our predecessors were not using certain methods before a certain point because those methods did not exist at that time. But there are many questions about what they actually were doing in each case, and how it compares to what we do for those same topics or related ones. We need not be motivated by an effort to defeat the general pessimist about scientific theories – he is already defeated, or at least will be by the end of this chapter. We are free to be motivated by the intrinsic interest of the relation between reliability and method.

Even if better methods are available today, are they being used, and used well? And is our confidence actually tailored to their reliability? And even if better general methods are available today might they not, for substantive reasons, have no added reliability value for a given topic of investigation? In such a case the calibration imperative would counsel no more confidence in our conclusions than we think our predecessors had a right to. The task of determining the reliability or comparative reliability of our methods looks more difficult the more subject-specific the method under consideration is, since usually one would need already to know or make assumptions about the subject to know whether such a method is generally any good. (Although if we cannot make *any* such assumptions, we are back to the basic problem of induction.) Content-independent arguments like those given above about general method go the distance for these methods.

There are several general reasons why this isn't a problem. One is, as explained, that now that we have shown that method is relevant to reliability, the burden is on the pessimist, not us, to show in particular cases that our methods, of whatever sort, are no better than those of our predecessors, if he wants to cast doubt on our beliefs in particular theories. And it is confidence in particular theories – Quantum Mechanics – not some general claim about such confidence in all of our theories, that the optimist has the goal of maintaining. Another is that what matters here is the comparison to our predecessors, not the absolute level of our reliability, for the question is whether their reliability problems are problems for us. This requires argument, but it is one that can be made.

We can see how the argument works with methods that require substantive assumptions by considering four cases, a two-by-two matrix over sameness and difference of substantive assumptions and method. In the first case we share substantive assumptions with our predecessors and our overall method is the same too, partly in consequence. In this case if we think they were wrong, then we will think we're wrong too, because on assumption of equally competent application we have the same conclusions! But that means those results will not be the matters at issue that we are currently confident about. (We also do not actually have the same conclusions so are unlikely to be in this case.) In the second case our substantive assumptions are the same and our method is different (which must be due to other aspects of it). This is the case we dealt with above: method is in general relevant to reliability, so this is material for a cross induction and moves the burden of proof to the pessimist.

In the third case our substantive assumptions are different but the overall method they are part of is the same as that of our predecessors. Unlikely, but in any case if our overall method is the same while the substantive assumptions are different, there must be some non-substantive part of the method that is different, and the central argument above applies. Once again the ball is in the pessimist's court. If our substantive assumptions are different and our overall method is, partly thereby, different, then since difference in method undermines the pessimistic induction, the burden of proof is with the pessimist. The fact that methods often involve substantive assumptions particular to a domain or a question, does not pose a problem for our response to the pessimist.

The pessimist may reasonably protest that even if we show our methods to be different from our predecessors', many of our predecessors also had methods that were different from their predecessors'. A lot of good it did them since they ended up wrong a lot of the time. This won't be enough to revive our naysayer's argument, though. The fact that our predecessors often had different methods from their predecessors' is material for undermining the induction over their predecessors to *them*. Thus, our predecessors' predecessors' failures did not render our predecessors' confidence unjustified either. So too for us, by the only induction that is left to make. This may also be why we tend to think of those of our predecessors who used methods that are a subset of ours as being in some real sense justified even when we also think their conclusions were wrong.

One might spy another negative induction in the offing here: it is true that our predecessors had different methods than their predecessors but they didn't succeed any better? Does this not provide inductive evidence that method is *not* relevant to reliability, which was a crucial premise of our cross induction?[16] This line is also doomed, for we showed non-inductively that method *is* relevant to reliability, even if methods involve substantive assumptions. Thus, there is something we know, relevant to the property this argument is inducing to – namely, the irrelevance of method to reliability – that is (maximally) relevant to whether that property can be expected to be there in the yet unexamined cases. This induction too is undermined by a cross induction.

The argument of this chapter also stands against a new pessimistic induction developed recently by Kyle Stanford (2006; see also 2000a, 2000b, 2003). The inference he describes involves conceivable alternate theories for explaining our evidence. Stanford points out that there were alternative possibilities to our predecessors' theories that were conceivable, and that showed our predecessors' theories false, but which they did not conceive of. We know these possibilities were conceivable, because we have since conceived them. There is no reason to think we are different from them in this respect; there must also be conceivables we have not conceived. And, roughly, we can't show that they don't undermine our theories because we're not conceiving of them.

Stanford focuses not on the challenge posed by beliefs of a certain sort, those about unobservables, but on ways at arriving at beliefs in science, in particular whether we can eliminate alternative explanations of phenomena. Since method is directly relevant to this, the argument of this chapter is relevant too. To perform Stanford's new induction, the cases of our predecessors and ourselves must be relevantly similar in the basis property, which here is proposed to be the affliction by unconceived conceivables relevant to the theoretical question the scientists in each case are investigating. This property is second-order and affects our predecessors' reliability, because as we know their affliction was partly responsible for their believing false theories. In fact their ways of dealing with the problem left them unreliable, as we can see because their theories were false. Since we are afflicted by the same problem, we can expect that we are unreliable. However, granting that we also are subject to unconceived conceivables, the question is whether our predecessors' unreliability that was due to this property really says anything about us.[17] For it to do so there must not be any other properties to show that we may be different from those predecessors in our reliability when facing possibility spaces of alternative theories.

Unfortunately for the pessimist, there have been a lot of discoveries of new methods for dealing with large theoretical possibility spaces. In particular, before the early twentieth century possible alternative explanations for phenomena and experiments were ruled out seriatim as they were conceived of. The early twentieth century saw the blossoming of a first round of techniques for ruling out large classes of theories without *conceiving* of their members (Roush, 2005: 218–223), Since then, of course, there have arisen more, and more sophisticated techniques.

Once we point out that our methods for dealing with unconceived conceivables are different in a way relevant to whether our theories are likely to be true, it is the pessimist's burden to show that despite these differences in method we are no more reliable than our forebears. The question is not, as Stanford sometimes suggests (Stanford, 2006: 131, 133), whether we have a method good enough to rule out all possible alternative theories. Since the pessimist is the one doing an induction, the question is whether our predecessors' faults should make us think we are in the same boat, so that we should dial down confidence in our particular theories. He has not shown this.

Thinking carefully about the pessimistic induction over the history of scientific failures shows that it must be a meta-induction if it is to make past failures *relevant* to us. The Preface Paradox is no paradox at the level where the optimist resides, with first-order confidence in particular theories. The pessimist must appeal not merely to the falsity of our predecessors' theories but to the unreliability of their ways of coming to their beliefs (as confirmed by their repeated false conclusions). But even if we grant their unreliability, nothing follows from this about whether we have a right to our confidence in our particular theories unless it is shown (1) that their unreliability is a reason to think we are unreliable and (2) that our believing we are unreliable forces a revision of our first-order beliefs on us. I have argued via our desire for calibration that (2) is true. However, (1) has not been shown, since the manifest difference in methods between us and our predecessors undermines the legitimacy of the pessimist's induction. The pessimist seems to be in even worse shape than having the burden of proof unloaded on him, since it is unclear how to show that method is not relevant to reliability without appealing to the general problem of induction.

Notes

1. This chapter has benefited greatly from discussions with Fabrizio Cariani and members of the Philosophy Department of the University of Washington, especially Arthur Fine, Bill Talbott, Andrea Woody, and Alison Wylie.
2. This work has been supported by NSF grant 0823418.
3. Recent efforts of this sort can be found in Leplin (1997), who argues for a link between novel predictive success and truth, and Psillos (1999), who argues for a link between referential continuity and descriptive accuracy.
4. This stance is of course somewhat similar to Arthur Fine's 'natural ontological attitude' (Fine, 1996). However, besides my focusing more on epistemology rather than ontology, there is also the difference that here it is a starting point against which, I argue, there is a compelling form of argument for ascent from our object-level beliefs to reflection on them, and from there descent to withdrawal of confidence in particular theories. It's just that, as I show below, it does not appear possible to make such an argument successful.
5. The claim I intend here does not assume as much externalist epistemology as it may seem. I grant that the optimist needs to have an ability to give reasons for her beliefs in theories, but the believer I have described can do this by citing her evidence for them. My claim that she need not have an argument ready for just any challenge to her right to her beliefs in theories is consistent with an obligation to take seriously any prima facie good argument against that right if she is to retain her status as justified. My point is that whether there are good arguments against her right is something that needs investigation before she loses her justified status by not having responses to them. Canvassing the space of such arguments is what I do in this chapter.
6. I am making a crude 'one drop' assumption about falsification here. But the pessimist must also be assuming the past theories are false or his negative evidence from the history of science disappears, since the property he is inducing to is 'probably false'. We can use here whatever sense of 'falsification' he is using there, and make the same argument. The pessimist could instead be arguing that past theories are false in virtue of being inconsistent with our theories, but that is not a first-order option in which the evidence for past theories is pessimistically relevant to our theories, which is the form of pessimistic argument considered in this section. Any second-order argument in which the falsity of their theories is supposed to be a reason to reduce confidence in our theories would fall victim to the central argument of this chapter, which follows.
7. If the p's are not independent, then there are conformably fewer sources of potential error than the number of p's, and the case is like a conjunction with fewer conjuncts. The comparison of few and many conjuncts in the preface situation is treated below.
8. Thus, a maximal level of justification is closed under conjunction. What is not closed under conjunction is rational *degree* of justified belief. The failure to distinguish these two facts, and the intuitions that correspond to them, causes much confusion about these kinds of cases.
9. This is also one problem that afflicts Andy Egan and Adam Elga's (2005) argument that one can't coherently believe one is unreliable without withdrawing

the beliefs in question. There fallibility of a given degree has simply been formulated as with the original Preface Paradox here, as the disjunction of the negations of the propositions I believe, which automatically cancels that same degree of confidence in the claims in question. The second-order formulation of fallibility I give and the solution I just described provide two independent resolutions of their problem.

10. This is not the only way to express reliability, and hence fallibility, though I think that any adequate way of expressing it as a property must be second-order, not first-order. I have switched to considering the reliability of full belief for ease of exposition. The same argument can be formulated for rate of success for a given level of partial belief in q.
11. One might be worried about the lower end of the spectrum of degrees between 0 and 1. What could it mean to say that someone was 20 percent reliable? It means, simply, that they get it right 20 percent of the time. They wouldn't be a very useful source (except in reverse, in a yes–no query), but that isn't the problem here.
12. One might notice that the second claim is the denial of a Type II defeater – a claim not that the purported evidence claims are false but that the purported support relation between the evidence and the hypothesis is weaker than thought. And one may wonder whether we don't need something higher than first-order probabilistic relevance to model that relevance. If we need second-order claims and these are intuitively relevant to the truth of Special Relativity, then we don't need any special explanation of why fallibility claims are relevant too. But first-order relevance accounts for (first-order) Type II defeaters handily. C is a Type II defeater of the claim that A supports B iff $\Pr(B/A) > \Pr(B/A.C)$. A Type II defeater *screens off* the relevance of A to B. The relevance of all the specific claims to the hypothesis that the speed of light is constant over reference frames can be represented at the level of first-order probabilistic coherence. The relevance of claims about fallibility cannot.
13. The obligation to *keep* yourself calibrated (as opposed to being in that state) can be formulated as a rule for updating our beliefs on discovery of information about our reliability level, so: $\Pr_f(q) = \Pr_i(q/[\Pr_i(q) = x.\ PR(q/\Pr_i(q) = x) = y]) = y$. That is, the final degree of belief you ought to have in q given that your current degree of belief in q is x and that you think the objective probability of q when you believe q to degree x is y, is y. So, for example, if you know that in the past when you responded to the weatherman saying there was a 30 percent chance of rain on a given day by having a 30 percent confidence in rain that day, and it actually rained 60 percent of the days, then you should update today's weatherman-induced 30 percent confidence in rain to 60 percent. This scheme is an extension of the Bayesian constraints on rationality, which relaxes certain idealizations in order to accommodate the fact that while it is unfortunate to discover you are less reliable than you expected, the only irrationality would be to fail to respond to that discovery. I am currently developing the details of this rationality constraint. Defense of the formal constraint is a very complicated matter, but my ground for endorsing calibration here is the empirical evidence that it is beneficial.
14. I am grateful to Arthur Fine for this objection.

15. There is a serious problem in this direction, first noticed by Hume, that comes from the potential error added in moves where we try to catch our errors. See Vickers (2000). I am able to escape it here because the situation allows me to make felicitously comparative judgements.
16. I owe this objection to Bill Talbott.
17. I say the property in question is reliability. We could start with an argument where our predecessors faced unconceived conceivables and ending up with *false* theories, but we already know that such an induction is undermined by our having different evidence sets from them.

References

Egan, A. and A. Elga (2005) 'I Can't Believe I'm Stupid', *Philosophical Perspectives*, 19(1): 77–93.

Fine, A. (1996) *The Shaky Game: Einstein, Realism, and the Quantum Theory*, 2nd edn (Chicago: University of Chicago Press).

Juhl, C. F. (1994) 'The Speed-Optimality of Reichenbach's Straight Rule of Induction', *British Journal for the Philosophy of Science*, 45: 857–863.

Kitcher, P. (2001a) *Science, Truth, and Democracy* (New York: Oxford University Press).

Kitcher, P. (2001b) 'Real Realism: the Galilean Strategy', *Philosophical Review*, 110: 151–197.

Laudan, L. (1981) 'A Confutation of Convergent Realism', *Philosophy of Science*, 48: 19–49.

Leplin, J. (1997) *A Novel Defense of Scientific Realism* (New York: Oxford University Press).

Psillos, S. (1999) *Scientific Realism: How Science Tracks Truth* (New York: Routledge).

Reichenbach, H. (1949) *The Theory of Probability* (Berkeley: University of California Press).

Roush, S. (2005) *Tracking Truth: Knowledge, Evidence, and Science* (Oxford: Oxford University Press).

Salmon, W. C. (1967) *The Foundations of Scientific Inference* (Pittsburgh: University of Pittsburgh Press).

Stanford, P. K. (2000a) 'An Anti-Realist Explanation of the Success of Science', *Philosophy of Science*, 67: 266–284.

Stanford, P. K. (2000b) 'Refusing the Devil's Bargain: What Kind of Underdetermination Should We Take Seriously?', *Philosophy of Science*, 48 (Supplement): S1–S12.

Stanford, P. K. (2003) 'Pyrrhic Victories for Scientific Realism', *Journal of Philosophy*, 100: 553–572.

Stanford, P. K. (2006) *Exceeding our Grasp: Science, History, and the Problem of Unconceived Alternatives* (New York: Oxford University Press).

Tenney, E. R., R. J. MacCoun, B. A. Spellman, and R. Hastie (2007) 'Calibration Trumps Confidence as Basis for Witness Credibility', *Psychological Science*, 18: 46–50.

Vickers, J. (2000) 'I Believe it, but Soon I'll Not Believe it Anymore: Skepticism, Empiricism, and Reflection', *Synthese*, 124: 155–174.

3
Metaphysics between the Sciences and Philosophies of Science

Anjan Chakravartty

1. The long birth of the modern sciences

In its conception, one of the motivational goals of this volume was to consider recent developments in the philosophy of science and to offer, with as much audacity as we could muster, some thoughts as to where these debates are going. In the following I will attempt to honor this motivation by tackling an audaciously large issue I see coloring the background of most contemporary work in general philosophy of science. For most of its history there has been a persistent antipathy in this area towards speculative metaphysics, but for several years now, a number of philosophers of science have been embracing it with increasing enthusiasm. These are starkly opposed approaches to the philosophical consideration of scientific knowledge, and something of an impasse exists between them; neither side has grappled very effectively with the other. In this chapter I aim to consider why current work on scientific knowledge has come to exemplify this divergence, and whether it is resolvable. While I cannot pretend to know with any certainty where the philosophy of science will go from here on this issue, it will be my contention that we should expect such a resolution to be unachievable, for reasons of philosophical principle.

Let me begin with an historical allusion. In the seventeenth century, natural philosophy, a human inquiry into the nature of the world along the lines of what we would now regard as several separate dimensions – philosophical, scientific, and theological dimensions, *inter alia* – began the slow process of resolving itself into inquiries that we now take to have separate identities. And so it is that now, both scientists and many philosophers look back to this early modern period as the time during which the modern incarnations of their disciplines began to take shape, slowly emerging from the fantastic amalgam that was natural philosophy.

Even the historically unschooled can appreciate, I think, the fascinating historiographical puzzles that hang over the seventeenth century. As Christoph Lüthy (2000: 164, 173–174) points out, historians of philosophy and historians of the sciences, in crafting their respective narratives of the modern origins of their disciplines, often point to exactly the same people and texts. Given that we now see these disciplines as engaging in distinct forms of inquiry, how is a historian to make sense of the overlap? While it is true that, in the seventeenth century, natural philosophy began increasingly to embrace empirical and experimental methods, which are often regarded as hallmarks of modern science, this actually led to more confusion than resolution so far as the historical story is concerned. As Lüthy (2000: 165) puts it: 'The confusing consequence was that everyone who was concerned with some aspect or other of the natural world came to consider himself *ipso facto* a "philosopher".' Early modern physicists and chemists called themselves 'experimental philosophers'. Telescopes and microscopes were philosophical instruments. Galileo's title was 'court philosopher'.

There are, of course, reasons to expect the story of how the sciences became inquiries apart from philosophy to be complicated. For one thing, the term 'natural philosophy' is often used to cover an enormous amount of ground, including essentially speculative and contemplative investigations focusing on puzzles associated with the Aristotelian tradition, as well as more empirical and experimental investigations associated with medicine, alchemy, metallurgy, optics, and astronomy. Another complicating factor is the ways in which philosophical concerns sometimes insinuate themselves, sneakily, into what we regard as scientific work even when they are supposed to have been exorcised. Margaret Osler (1996) argues, for example, that the common view that the rise of the mechanical philosophy in the seventeenth century banished the scholastic idea of final causes is mistaken. Final causes, she maintains, were simply reinterpreted so as to fit in with the mechanical philosophy. In a similar vein, George MacDonald Ross (1998) documents the persistent appeal to 'occult' qualities during this period.

Having raised these historiographical issues, it is not my intention to resolve them. I mention them because I think they foreshadow very effectively the philosophical problem that is my real interest here. However the story of how the modern sciences arose from natural philosophy is told, there is at least one thing that most people agree on, and that is the idea that in the process, the sciences became independent of many of the philosophical and other preoccupations that had previously informed theorizing about the natural world. For example, where once it made sense to think about issues concerning divine providence and the nature of the soul as relevant to inquiries into the nature of atoms, or space and time, that is no longer the case. Clear examples of what we would now regard as philosophical and theological issues are no longer entertained as relevant to scientific inquiry. That is not to say that there is any precise definition of what the sciences are that might clearly and unproblematically demarcate them from other pursuits. Branches of the sciences differ a great deal from one another with respect to canonical markers, such as how mathematical or experimental they are, whether they make novel predictions as opposed to merely accommodating the data, and so on. But leaving annoying demarcation issues to one side, it is clear that we now have separate domains, and no one doing physics today is expected to worry about whether God could have created the entire contents of the universe just as it is, but ten meters to the left.

To a great extent, then, it is reasonable to say that the sciences today are independent of the major philosophical and other preoccupations that were previously part and parcel of theorizing about the natural world. I say 'to a great extent', because for the remainder of this chapter, I will be investigating what I take to be an exception, viz. metaphysics. I will argue that whatever philosophical baggage the sciences have left aside en route to the present, a certain degree of what is reasonably called metaphysics remains and cannot be removed. Given what I mean by 'metaphysics', this conclusion may or may not prove especially controversial. I will argue further, however, that there is no correct answer *per se* to the question of what *degree* of metaphysical speculation is properly considered reasonable in interpreting scientific knowledge. In arguing for this further claim, I will suggest that ultimately, we are left with an irresolvable methodological relativism concerning what degree of speculation is appropriate to philosophical analysis in this context. It is this lack of absolute conviction in determining just how metaphysical scientific knowledge is that I take to parallel the historiographical uncertainty surrounding whether early modern natural philosophy is science or philosophy.

2. Metaphysics in science: for and against

As a precursor to arguing for these theses, perhaps it would be useful to say something about what metaphysics *is*. Let us begin with the textbook definition. As one learns on the first page, the term itself originates from the classification of Aristotle's works three centuries after his death by Andronicus of Rhodes, the works placed after the *Physics* being labeled, naturally, *Metaphysics*. The inquiry itself was identified chiefly with two things: the study of being *qua* being, or the most general nature of existence and the natures of things that exist; and the study of first causes and theology (in Aristotle, the ultimate cause being God, the unmoved mover), which contributes the connotation of providing fundamental or basic explanations. It has a significant a priori dimension, aiming to provide an explanatory account of observable phenomena in terms of underlying realities. This a priori character is manifested in the ways in which its arguments typically proceed, with emphasis on conceptual analysis, intuitions, and explanatory power *inter alia*. Metaphysics as practiced since ancient times through to the present has retained much of the flavor of this original conception. I will leave aside the idea of providing fundamental explanations here, for although many of the great systematic philosophers aspired to do precisely that, metaphysics often includes less ambitious theorizing. The idea of a significantly a priori investigation into something explanatory that underlies the observable phenomena, however, is central.

Some may worry, perhaps, that the emphasis I have just given to the distinction between observable aspects of the world and things underlying them furnishes a misleading characterization of the metaphysical enterprise. The implication here is that 'things underlying' the observable phenomena are not themselves observable, strictly speaking. 'Observable' in this context is used in the philosopher's sense, to refer to those aspects of reality that are in principle detectable using only human sensory faculties. Unobservable entities are thus things that are not in this way detectable. Some may view the explicit invocation of this distinction as tangential to the more general and simply stated aim of metaphysics: that of considering the natures of things in the world. I submit, however, that there is a very short distance between considering natures and speculations regarding the unobservable. In providing accounts of the natures of various aspects of the world, be they observable or unobservable, metaphysicians have traditionally appealed to strictly unobservable elements underlying their surface appearances – universals, substances, bundles, necessities, possibilia, and so on. Given the immediacy of such speculation in carrying out the generic aim of metaphysics, it seems reasonable to take it as constitutive, and I will do so in what follows.

With this first definition of metaphysics in mind, let me now pose the central question of this chapter: do the modern sciences involve or incorporate metaphysics, in just the way that early modern natural philosophy involved and incorporated other philosophical preoccupations? Some say yes and others no, and my own answer to this question will build on these responses, so let us briefly consider them in turn.

In recent memory, the most strident denunciation of metaphysics in connection with science has come from logical positivism, the founding movement of the philosophy of science as a self-aware, philosophical discipline; it began with a commitment to the positivist premise that, in the sciences, human thought has finally progressed beyond its prior religious and metaphysical modes. The emphasis placed by the positivists and later logical empiricists on sensory experience as the subject matter of the sciences was intended precisely to excise any hint of metaphysics from our understanding of scientific knowledge. More recently, Bas van Fraassen has developed his own version of empiricism in the scientific context, constructive empiricism, in hopes of carrying the empiricist baton forward. In both cases, the idea of an investigation into things underlying sensory experience which might explain aspects of the observable, a priori or otherwise, is an anathema.

Some qualification is necessary here, for clearly these rejections of metaphysics are not so naïve as to miss the fact that much scientific work and discourse makes use of terms whose intended referents appear to be unobservable entities and processes. The empiricist point here is not that these surface features of science are illusory, but rather that they should not be interpreted as leading to knowledge of the unobservable. Thus, it may well be that scientists routinely theorize, experiment, and communicate with one another in a way that suggests speculations about the unobservable, but this, according to various empiricist philosophers of science, should not be taken to reflect the true aims or nature of science. Such suggestive speculation, where it takes places, must be understood in a particular way, as simply serving to further and systematize our claims about the observable. It is a means to an end, where both are conceived in empiricist terms. In this way, the empiricist and the metaphysician may offer precisely the same sorts of descriptions of the surface features of scientific investigation. The difference between them concerns how these surface features are to be interpreted.

How successful, then, are these rejections of metaphysics? Neither has been uncontroversial. Logical positivism and empiricism faced the accusation that the verifiability criterion of meaning, according to which putatively factual propositions amount to meaningless metaphysics unless there are empirical procedures by which they can be verified or confirmed, was itself a piece of metaphysics, or followed from metaphysical commitments (see Alston, 1954, for instance). In Carnap's later work, the linguistic frameworks within which scientific theories are expressed comprise analytic principles which (partially) constitute the meanings of theoretical terms, and determine what counts as confirming evidence for the framework as a whole. The choice of a linguistic framework cannot be made on the basis of factual evidence, however, for there is simply no framework-transcendent conception of evidence on the basis of which to make such a choice. Recent commentators such as Michael Friedman (1999) have argued that this is nothing less than a neo-Kantian account of the a priori. Constructive empiricism, on the other hand, faced the accusation that its claim that science aims to produce theories that are empirically adequate, where empirical adequacy is to be understood in terms of truth concerning the observable, makes metaphysical commitments in extrapolating beyond the empirical data, and in appealing to the modality inherent in the notion of observability.

These are all important challenges to the rejection of metaphysics in the context of scientific knowledge, but I will not dwell on them here. Let me rather point out another, more general worry attaching to empiricist hostilities to metaphysics, whatever their form. *Prima facie*, the sciences are *plainly* metaphysical, insofar as they appear to take a very strong interest in phenomena underlying the observable. The sciences routinely take an interest, apparently, in particles and fields and genes and enzymes and an endless parade of entities that cannot be detected with the unaided senses. Wherever possible, hypotheses concerning these entities are subjected to empirical tests, but tests never render them *observable*, and some hypotheses are accepted for primarily explanatory reasons. As a consequence, in the attempt to rid scientific practices of the semblance of metaphysics, empiricisms are driven to extraordinary lengths. The logical positivists, for example, were driven to a non-realist semantics, according to which all terms putatively referring to unobservables are meaningless unless reinterpreted in terms of observables – a hopeless semantic theory which led ultimately to the downfall of positivism itself. Constructive empiricism adopts a realist semantics, but in remaining agnostic about the existence of so many entities apparently described by seemingly empirically adequate theories, it too offers an interpretation of scientific claims that is a very long way from what they say on their face. Empiricisms can only make sense of the sciences by radically reinterpreting what they plainly seem to be doing, and that may well make one suspicious.

What, then, of the opposing view, that the sciences incorporate metaphysics as a matter of course? Many post-positivist philosophers of science have suggested this, but again, not uncontroversially. Edwin Burtt (1959/1925) maintained that not only are metaphysical concepts such as substance, essence, and form pivotal in the work of Copernicus and Kepler, but that in Newton, the concepts of space, time, and mass function in much the same way. R. G. Collingwood (1998/1940) discussed what he took to be metaphysical assumptions underlying physics from Newton's mechanics to Einstein's theories of relativity. Thomas Kuhn (1962) argued that metaphysical commitments, such as commitments to teleological versus mechanistic explanations, are one of the ingredients constituting the disciplinary matrix that governs a period of normal science. Again, there are criticisms one might press in connection with these particular views as they pertain to the modern sciences, but let me mention just one worry that will attach to any such position. It is not always entirely clear how to draw the line, or indeed, whether it is possible to draw a line, between the aspects of theories these commentators identify as metaphysical, and aspects that are properly considered empirical. Given this, the substance of the claim that scientific theorizing incorporates elements of metaphysics may seem inevitably disputable.

Consider an example: in the sociology of scientific knowledge, it is not uncommon to find the suggestion that metaphysical presuppositions (the adoption of which, it is claimed, is best analyzed in terms of social, economic, and political causes) play a role in what ends up as scientific fact. Barnes et al. (1996: chapter 2), for example, examine Millikan's celebrated oil-drop experiments conducted in the early twentieth century, the goal of which was to investigate the hypothesis of a fundamental unit of electric charge. Millikan suspended ionized droplets of oil between two charged plates, and using Stokes equation, measured their charge, ultimately concluding that all of the results were integral multiples of (approximately) 1.602×10^{-19} C – the fundamental unit. Interestingly, a rival physicist, Ehrenhaft, conducted similar experiments at around the same time at the University of Vienna, using tiny metal spheres instead of oil, and his results indicated ever smaller measurements of electric charge, without any suggestion of a fundamental unit as such. The authors suggest that Millikan reached his own conclusion because he was operating with the assumption that electric charge *must be* quantized, an assumption licensed by a physics community excited by the promise of the quantum hypothesis more generally, where previously the idea of a continuous electric field or fluid had been in vogue. Different interpretations of Millikan's data were clearly possible, they argue, citing historian Gerald Holton's analysis of Millikan's notebooks, indicating the selective use of some data and the laying aside of others. His metaphysical assumption regarding quantization drove these decisions, they claim. But was the quantum hypothesis really a metaphysical assumption influencing Millikan's work, or was it rather an empirical hypothesis established by his experiment, and indeed, other experiments in physics at the turn of the century?

The question of whether scientific hypotheses regarding unobservable entities and processes are metaphysical or empirical is inherently messy. Consider another example. In a recent study of the development of cancer research, J. A. Marcum (2005) argues that two conflicting metaphysical presuppositions – reductionism and organicism – have shaped scientific investigation into the nature of cancer in recent decades, and have themselves been shaped by this research in turn. Reductionism here takes the form of genetic determinism, the idea that certain biological states and processes can be explained in terms of genes, and organicism is the view that emergent phenomena at higher levels of biological organization are crucial to such explanations, such as the explanation of cancer cell production as a consequence of abnormal tissue organization. But again one might ask: are these competing biological 'isms' really metaphysical presuppositions, or are they empirical hypotheses to be tested in the course of cancer research? Given that, as Marcum suggests, they are amenable to reformulation directly informed by scientific research, does this suggest that they are in some sense empirical?

I have just offered considerations on both sides of the same question. On the one hand, among other worries, empiricist critiques of the idea of metaphysics in connection with the sciences may give the impression of rather severe and ideologically driven rational reconstructions; but on the other hand, among other concerns, it is not entirely clear that the sorts of examples found in the literature of metaphysics operating in modern scientific practice are really metaphysical after all. I believe that scientific knowledge is unavoidably metaphysical, but I also believe that the sorts of considerations I have just sketched are incapable of deciding the issue one way or the other. Understanding the relation of metaphysics to the sciences will require a different and more roundabout approach. Let us turn to this now.

3. Stances, rationality, and values

I have mentioned in passing some examples of people who contest the idea that metaphysics is relevant to the sciences, all of whom identify themselves as empiricists. Though one need not be an empiricist to make such claims, it is not surprising that the most adamant objections to metaphysics in the context of scientific knowledge come from this group. Empiricism is often presented in opposition to forms of metaphysical inquiry. It is too strong simply to say that there is a conflict between empiricism and metaphysics, since the rejection of certain kinds of metaphysical inquiry often proceeds on the basis of what are commonly described as metaphysical arguments. Humean austerity is sometimes referred to as arising from and constituting a Humean metaphysic, for example. But metaphysics as I have described it – an inquiry into the nature of things underlying the observable, having explanatory and a priori dimensions – is no doubt something at odds with most empiricisms. In order to examine the idea of metaphysical inquiry a little more precisely now, I will borrow a tool from some recent work by van Fraassen: the idea of an epistemic stance.[1]

An epistemic stance is one concerned with the production of knowledge, and I will use the term 'stance' in this sense henceforth. It is perhaps easiest to understand the concept of a stance in contrast to what we might call propositions, or claims regarding matters of putative fact. Matters of fact would seem to include the claims that grass is green, that chlorophyll facilitates photosynthesis, and that carbon dioxide is used in the process. If endorsed, the appropriate attitude towards such propositions is belief. A stance, on the other hand, is a cluster of attitudes, commitments, and strategies relevant to the production of factual beliefs. Unlike propositions, they do not make claims about the world, but determine how epistemic agents go about making claims about the world. Stances are not believed, but adopted, held, and expressed in human action. They may include beliefs, but unlike propositions, their relation to knowledge seekers is not exhausted by belief in any strictly propositional content.

The stance characteristic of those who do metaphysics, for example, involves taking seriously the idea of explaining observable phenomena in terms of things underlying them, as well as attempts to answer such demands by speculating about the unobservable. It may involve other things, but this much will suffice for present purposes, and I will use 'metaphysics' as a label for this stance, so identified, henceforth. In contrast, a stance characteristic of many empiricists is constituted (again, no doubt among other things) by a rejection of such demands and speculations. As noted earlier, these features of metaphysics annoy empiricists, but to this we can now add the observation that this annoyance is most economically understood at the level of stances. Rather than list the countless factual claims of which empiricists disapprove, whether they concern the existence and nature of universals or subatomic particles, it is most economical simply to note that metaphysics is a stance of which empiricists disapprove.

Regarding metaphysics and empiricism, the crucial questions for the current discussion are these: on what basis do we choose the stances associated with them; and on what basis do we assess the wisdom of adopting one as opposed to an apparent rival? Van Fraassen (2002) identifies two such grounds, and I am inclined to agree. The first is rationality: one should choose a stance that is rational, where this is defined in a broadly pragmatic way in terms of internal coherence. Accordingly, the 'defining hallmark' of irrationality is 'self-sabotage by one's own lights' (Van Fraassen, 2004: 184). In the absence of such a failing of one's epistemic project as assessed by one's own standards, the stance (or stances) associated with that project is rational. This is a permissive view of rationality, allowing that different and mutually incompatible stances may be rational, and it follows from such a view that there may be cases in which no one stance and resultant set of beliefs is compelled on grounds of rationality alone. The second consideration relevant to choosing stances is the idea of values. Beyond the constraint of rationality, the values of epistemic agents play a decisive role in determining what stances are appropriate for them. Values are agent-relative, and as a consequence, the adoption of stances is generally a relativistic proposition. Those with different values are at liberty, within the bounds of rationality, to choose differently in accordance with their own values.

Thus, an empiricist may reject metaphysics by committing to epistemic policies that are incompatible with it, but so long as metaphysicians themselves adopt rational epistemic stances, this cannot suffice to convince metaphysicians that they are ill-advised in their methods of inquiry. If rationality is the only constraint that applies uniformly to all agents adopting stances, and different, mutually incompatible stances are rational, then the framework for debate concerning one's choice of stances is subject to a form of methodological relativism. The metaphysician and the empiricist appear to be separated here by different intuitions, or values, regarding (at least) two centrally important matters. The first is the question of what sorts of phenomena stand in need of explanation; for what sorts would an explanation be desirable? In this regard, many empiricists are content to rest with the observable, and view the desire to extend belief to explanatory phenomena underlying the observable with evident disdain. On the other side, those adopting the metaphysical stance value such explanations, and thus naturally take a different view. Closely related to this difference is the question of what sorts of explanations are viewed by epistemic agents as obscure or unilluminating, as opposed to contributing genuine insight. Again, many empiricists view speculations regarding the unobservable negatively in this regard, while those of a metaphysical bent naturally see things differently.

Some will worry here that what I have suggested amounts to an alarming relativism, and that since alarming relativisms are to be resisted at all cost in view of the terrible epistemic consequences they entail, the idea of a stance relativism encompassing metaphysics and empiricism must be resisted too. But there is no cause for alarm here. While certain forms of relativism about knowledge are clearly pernicious, I believe the methodological relativism proposed here to be innocuous. Note, for instance, that it is not a relativism allowing different epistemic agents to adopt contrary beliefs regarding matters of putative fact. Where a metaphysician may be tempted to claim P, it is not the prerogative of the empiricist to claim not-P. Rather, the empiricist remains agnostic about the truth or falsity of P. Where the metaphysician may affirm the existence of quarks or possible worlds, the empiricist does not deny their existence, but remains agnostic. In this way the set of factual beliefs endorsed by the empiricist may be a subset of those endorsed by the metaphysician – they may share beliefs about observables, but the former (and not the latter) will refrain from affirming and denying metaphysical claims regarding unobservables. Thus, the form of relativism applicable here avoids the controversy of rationally sanctioned contradictions. It is this controversy that fuels debates about relativism, but it is of no interest in this case.[2]

With this digression on the nature of epistemic stances in hand, I believe we are now in a position to tackle the questions with which I began this chapter, concerning whether metaphysics plays a role in the modern sciences, and whether it properly has a role in the interpretation of modern scientific knowledge. In what follows, I will assume that the stances associated with metaphysics and the empiricisms that contest it are rational, and embrace the methodological relativism this appears to entail.

4. The sciences as inevitably metaphysical

I began this discussion by stating that I would argue for two conclusions. The first was that the sciences inevitably incorporate a degree of metaphysics, and the second was that there is no one, correct view as to how much metaphysics one should invoke in giving a philosophical account of scientific knowledge. In this section I will focus on the first of these claims. I will argue for it by suggesting that even on the most metaphysically austere, contemporary conception of the sciences – that offered by empiricist philosophies of science – there is a metaphysical dimension to this sort of inquiry. If this suggestion is compelling, it will follow as a matter of course that on other, less austere conceptions, metaphysical claims are part and parcel of scientific work.

Admittedly, some forms of empiricism need have no tolerance for the epistemic policies characteristic of speculative metaphysics. Take for example a strict phenomenalism of the present moment, according to which knowledge claims are restricted to those describing current sensations. In restricting itself so acutely, such a position, it would seem, need make no recourse to metaphysical speculation. It is an interesting question whether such a view could be developed to support a coherent epistemology, but for present purposes, we may leave this question to one side. Contemporary empiricist philosophies of science, as it happens, are not so stingy: they extend the boundaries of what is knowable in various ways beyond impressions and ideas. Most empiricist philosophers interested in the sciences today tacitly embrace some metaphysical speculation in order to preserve the coherence of their positions, or so I will contend.

For the sake of illustration, let me take the most influential, current example of this sort of approach to the sciences, van Fraassen's 'constructive' empiricism, as representative. This position endorses claims not merely about sensations, but about that which is observed and indeed, about that which is observable. In this way it is representative of contemporary empiricist philosophies of science more generally. They are not versions of idealism that deny the existence of an external world, nor are they versions of quietism with respect to the world beyond impressions and ideas. On the basis of empirical evidence, their epistemic grasp extends to a knowledge of observable entities and processes that exist quite independently of ideas. They aspire to some knowledge of a world that is external to human cognition but nonetheless the subject of experience.

As soon as an empiricist claims knowledge of more than sensations and ideas and includes observable entities and processes, however, her epistemological landscape changes very significantly in comparison to her phenomenalist cousin. In the former case but not the latter, it is paramount that one have an error theory. It is a commonplace to note that the more ambitious empiricist does not ultimately believe, for example, in the existence of 'objects' experienced in optical illusions and hallucinations, despite the fact that she clearly has sensations in connection with these events. Experiences are not always veridical in what they convey about things in the world external to cognitive processing. Not all observations are created equal; some lead to facts, others give misleading impressions, and yet others lead us into serious error. Through the mist clinging to the ice fall at Fox Glacier I see a large furry shape leap gracefully into a cavern of ice and disappear. Is it a play of the light, or the abominable snowman? In order to know something about observable things, not merely her sensations, the empiricist must have an understanding of what it means for some observations to be better than others, and how to differentiate the better from the worse. In making these judgements, we must also know how to describe our observations in terms of various categories of objects and events and their salient properties, in ways that allow us to communicate successfully with those with whom we interact, whether in scientific or everyday contexts. We must know how to draw implications from these experiences, as required by judgements of confirmation with respect to hypotheses and theories.

Again, the idea that empiricisms that endorse a knowledge of observable phenomena beyond mental phenomena require some sort of error theory is not new. Indeed, both the requirement of an error theory and the sorts of judgements this entails have been widely appreciated, not least by empiricists. Hence Wilfrid Sellars and others, for example, were keen to point out the 'myth of the given'. Facts about the world, they noted, are not things that can be simply or transparently read from raw experience, whatever that might be. In a similar vein, Norwood Russell Hanson, Thomas Kuhn, and Paul Feyerabend made interestingly related cases to the effect that scientific and everyday observations are theory laden, and this idea has long since been absorbed into the philosophy of science. What we see, they suggested, is importantly shaped by the cognitive and theoretical background we bring to observation. In Kuhn's terminology, this background inevitably includes the paradigmatic frameworks in which observation occurs, comprising exemplars for problem solving and metaphysical commitments regarding the various standards empirical investigation should meet.

So how does one distinguish veridical experiences from others about which one is more circumspect? How does one carve the referents of impressions and ideas into useful and communicable ontological categories, and extrapolate from these experiences in the ways required by scientific hypothesis and theory testing? The answers to these questions do not come from experience. They have their source in intersubjective understandings or traditions exemplified by a community, which furnish standard or customary answers to questions such as these. These understandings or traditions are generally tacit, but presumably need not be in all cases. They are shared by the members of an epistemic community, unifying their practices of empirical investigation. For Kuhn, the disciplinary matrix and exemplars that unify normal science under the rubric of a paradigm perform precisely this role. As van Fraassen (2002) puts it, one must know *how* to experience things before one can derive knowledge from experience.

Now, one might well ask, what sorts of things are these understandings, or traditions, or paradigms? They are unobservable, cognitive, cultural, heuristic entities, which underlie the phenomenon of observation on which empiricism is partially grounded. Like many complex social entities, we posit them and speculate about them for important explanatory reasons, to account for various of our experiences: in this case, the successful judgements and interactive work typical of scientific investigation. This positing and speculation is by no means gratuitous. Indeed, on the very empiricist views of science that are dubious of these practices, they would appear to be essential: they furnish empiricism with an error theory. In scientific contexts they explain how, in van Fraassen's terms, we manifest the capacity to learn things about the observable. Note, however, that the tacit understandings and conventions that inform scientific observation are things *we bring* to observation. They are not read off from experience, but are rather preconditions of experience, and thus have an a priori dimension. It is in this sense that even a very determined empiricist, it seems, must admit some metaphysics in connection with the sciences after all. On even the most metaphysical austere, contemporary conception of the sciences, they are inherently metaphysical.

To be sure, I have taken some liberties in describing how the sciences inevitably incorporate metaphysics, but none of these liberties undermines the point at issue. One might wonder, for instance, whether a social, intersubjective thing such as a tradition or a paradigm is properly described as an entity *per se*, let alone an unobservable one. If disciplines as disparate as physics and sociology teach us anything, however, they demonstrate that not all entities are amenable to neat ontological packaging, and the complexity of social entities and practices presents no obvious bar to a discussion of them in these terms. Neither are such understandings strictly observable, though aspects of them may be suited to study and detection. Those accustomed to applying the term 'a priori' to beliefs and propositions may balk at its use here in application to features of observation that are often tacit, but there would seem to be no obvious reason to restrict its use in this way. Others may wish to reserve the term for so-called analytic truths, and thus resist its application to things that are susceptible to change over time, but this too is a merely stipulated restriction. Still others may hope to disentangle the metaphysics from the a priori, and thus grant that there are a priori dimensions to scientific practice, while denying that they constitute any sort of metaphysics. But such disentangling, I submit, is impossible. The conventions we bring to the world of scientific experience make substantive contributions to our understandings of the natures of things that exist: they determine the ontological categories we employ and how we judge the coherence and confirmation of scientific claims. None of these conventions is read from experience – they are frameworks for experience itself.

5. The aims of science

Are the modern sciences devoid of metaphysics in the way that they are now free of certain other philosophical considerations that once infused natural philosophy? There is more to be said in affirming my negative answer to this question, but for the present, let me turn now to the second conclusion I hoped to reach in this chapter: the idea that there is no one or correct answer to the question of what degree of metaphysical speculation is properly considered essential to the interpretation of scientific knowledge. Different answers to this question have been championed by contemporary philosophers of science, with those expressing an antipathy towards metaphysics lined up on one side, and those who have been increasingly embracing it on the other. Here I will be brief, for it would appear that in just the way epistemic stances are susceptible to a form of methodological relativism, so too is the answer to this question.

Logical positivism engendered distaste for metaphysics within the philosophy of science with long-lasting influence, and van Fraassen's claim that the aim of science is not truth but empirical adequacy has also been influential, functioning as a slogan for anti-realists of various kinds concerning scientific knowledge. Those falling into these camps commonly subscribe either to a minimalist Humean metaphysic, or to quietism about metaphysical questions generally. In recent years, however, a resurgence in analytic metaphysics has been joined by a growing number of philosophers of science. Authors engaged in what is nowadays called the 'metaphysics of science', for example, commonly make claims to the effect that scientific knowledge simply cannot be properly interpreted unless one gives an account of the implicit features of this knowledge. These accounts typically include elaborate ontological commitments to things such as laws of nature, essences, and causal powers.[3]

What philosophical presuppositions separate these two approaches to science? Can the opposition between them be resolved? I submit that the approach to science that one thinks best reflects its true character is simply a reflection of one's prior adoption of particular epistemic stances such as metaphysics or empiricism. What degrees of speculation and concomitant distances from empirical investigation one thinks are appropriate for anything calling itself science, and what degrees and distances one thinks are appropriate in interpreting scientific knowledge, are generally a clear function of the epistemic stances one adopts. I have suggested that certain metaphysical assumptions may be inescapable in the practice and interpretation of what we call science. Here is another example, to illustrate the present point: an acceptance of some conception of the uniformity of nature, many have suggested, may be a precondition for practices such as scientific generalization and prediction regarding empirical phenomena. But what more precisely this conception amounts to – whether it is properly understood in a minimalist, Humean fashion in terms of brute regularities, or more elaborately in terms of laws and *de re* necessity – is something that cannot be determined except from the perspective of stances like metaphysics and empiricism. And as I have suggested, the adoption of these sorts of stances is susceptible to variable choice, itself subject to an indissoluble relativism.

It is arguable, of course, that this conclusion is overly hasty. Some may hold hopes of discovering the true nature of science, and thus how metaphysical it and the knowledge it produces is, simply by studying it more carefully as a practice. Perhaps if we were simply more careful and objective observers of the nature of scientific practice, we would find definitive answers there. But this, I suggest, is hopeless. Among scientists who care to think about these questions, there is no consensus regarding how empirical or speculative an investigation must or may be in order to count as science (consider recent disputes over the scientific status of string theory), and neither is there consensus regarding whether theories should be interpreted realistically, instrumentally, or otherwise. And even if there were majority opinions among scientists regarding these questions one way or the other, it is highly unlikely that taking a vote would yield the desired information. The sciences are human practices, and like any complex human practices, their nature and significance are open to interpretation. Look behind any particular interpretation here, and you will find the stances that condition it.

Ultimately, I suspect that there is little reason to suppose there *should* be any one, correct answer to the question of what degrees of metaphysical speculation are appropriate to the sciences, or to interpreting its products. Different epistemic agents inevitably draw the line in different places, and their choices are to a large extent conventional, made in accordance with standards that need not be shared among those who take an interest in the sciences as a subject of philosophical investigation. In just the way that historians may be rightly uncertain as to whether one should properly regard early modern natural philosophy as science or philosophy, I believe we should accept that just how metaphysical the modern sciences are, and how much of a metaphysician one needs to be in order to give an account of scientific knowledge, are questions open to different and ultimately indefeasible understandings of the aims of science.

Acknowledgements

This essay was written while visiting the Sydney Centre for the Foundations of Science, and aspects of it were presented at the Universities of Canterbury, Otago, Waikato, Melbourne, Macquarie University, and the annual meetings of the Canadian Society for the History and Philosophy of Science and the British Society for the Philosophy of Science. I am grateful to colleagues at these places for excellent feedback, and to Mark Colyvan and Philip Catton for making the antipodean adventure possible. The current volume provided a welcome opportunity to draw some broader consequences from a major focus of my work in recent years, and for this I would like to thank Jacob Busch and P. D. Magnus.

Notes

1. Due to constraints of space, the foregoing is a highly telescopic summary of the nature of stances, rationality, and values in connection with metaphysics and empiricism. For details spelling out and defending some of the assumptions made in this section, see Chakravartty (2007). For more background in support of the argument of section 4, see Chakravartty (forthcoming).
2. Consequently, I would distinguish the methodological relativism discussed here from relativisms associated with some approaches to the sociology of science and standpoint (and related) theory in feminist studies of science, both of which are embroiled in precisely this sort of controversy. Recent work in

epistemology also considers the idea that some beliefs may be rationally permissible but not obligatory, leading to rational differences in doxastic attitudes towards certain propositions. Cf. Rosen (2001), who affirms this in connection with nominalism and realism about abstract entities. White (2005) and Christensen (2007) both argue against the idea, but the arguments appear to leave untouched the present discussion: they consider only evidential reasons for belief, thereby ignoring non-evidential reasons including values as considered here; the arguments turn on cases of rationally sanctioned contradictions, thereby ignoring the contrast between belief and agnosticism paramount here; they employ emotive examples as intuition pumps (jurors placidly accepting contrary opinions while deliberating about a terrible crime, etc.), that seem inappropriate to the present context.

3. For just two prominent examples of this growing literature, see Ellis (2001) and Bird (2007). Cf. Psillos (2005), who argues that science does not imply a non-Humean conception of reality, and that debates between Humean and non-Humean conceptions are independent of other epistemological commitments concerning the sciences, such as scientific realism.

References

Alston, W. P. (1954) 'Are Positivists Metaphysicians?', *Philosophical Review*, 63: 43–57.

Barnes, B., D. Bloor, and J. Henry (1996) *Scientific Knowledge: a Sociological Analysis* (Chicago: University of Chicago Press).

Bird, A. (2007) *Nature's Metaphysics: Laws and Properties* (Oxford: Clarendon Press).

Burtt, E. A. (1959/1925) *The Metaphysical Foundations of Modern Physical Science: a Historical and Critical Essay* (London: Routledge & Kegan Paul).

Chakravartty, A. (2007) 'Six Degrees of Speculation: Metaphysics in Empirical Contexts', in B. Monton (ed.), *Images of Empiricism: Essays on Science and Stances, with a Reply from Bas C. van Fraassen* (Oxford: Oxford University Press).

Chakravartty, A. (forthcoming) 'A Puzzle about Voluntarism about Rational Epistemic Stances', *Synthese*.

Christensen, D. (2007) 'Epistemology of Disagreement: the Good News', *Philosophical Review*, 116: 187–217.

Collingwood, R. G. (1998/1940) *An Essay on Metaphysics*, ed. R. Martin (Oxford: Clarendon Press).

Ellis, B. (2001) *Scientific Essentialism* (Cambridge: Cambridge University Press).

Friedman, M. (1999) *Reconsidering Logical Positivism* (Cambridge: Cambridge University Press).

Kuhn, T. S. (1962) *The Structure of Scientific Revolutions* (Chicago: University of Chicago Press).

Lüthy, C. (2000) 'What to do with Seventeenth-Century Natural Philosophy? A Taxonomic Problem', *Perspectives on Science*, 8: 164–195.

Marcum, J. A. (2005) 'Metaphysical Presuppositions and Scientific Practices: Reductionism and Organicism in Cancer Research', *International Studies in the Philosophy of Science*, 19: 31–45.

Osler, M. J. (1996) 'From Immanent Natures to Nature as Artifice: the Reinterpretation of Final Causes in Seventeenth-Century Natural Philosophy', *The Monist*, 79: 388–407.

Psillos, S. (2005) 'Scientific Realism and Metaphysics', *Ratio*, 18: 385–404.
Rosen, G. (2001) 'Nominalism, Naturalism, Epistemic Relativism', *Philosophical Perspectives*, 15: 69–91.
Ross, G. M. (1998) 'Okkulte Strömungen im 17. Jahrhundert', trans. A. Beriger, in J.-P. Schobinger (ed.), *Friedrich Ueberwegs Grundriss der Geschichte der Philosophie, Reihe 5, 17. Jahrhundert, Band 1* (Basel: Schwabe).
van Fraassen, B. C. (2002) *The Empirical Stance* (New Haven: Yale University Press).
van Fraassen, B. C. (2004) 'Replies to Discussion on *The Empirical Stance*', *Philosophical Studies*, 121: 171–192.
White, R. (2005) 'Epistemic Permissiveness', *Philosophical Perspectives*, 19: 445–459.

4
Nominalism and Inductive Generalizations

Jessica Pfeifer

In this chapter, I argue that there must be properties[1] that are not merely sets, classes, pluralities, or mereological sums of concrete particulars, whether concrete particulars in this world or those in all possible worlds. Otherwise, we would be unable to formulate inductive generalizations. In fact, our ability to formulate inductive generalizations precludes almost all forms of nominalism. However, my main focus is to establish that class nominalism is untenable. Once this is established, I will argue that most versions of nominalism suffer from the same fundamental problem.

1.

I use the term 'nominalism' to refer to accounts of properties that countenance only concrete particulars and sets or classes[2] of these.[3] The concrete particulars might be limited to those in the actual world or include those in other possible worlds. I will use 'individual', 'object', and 'particular' interchangeably with 'concrete particular'. When discussing abstract particulars, I will use the terms 'abstract particular' or 'trope'. 'Nominalism' is sometimes used in a narrower sense, following Quine, to refer to accounts that do not posit *any* abstract objects, including sets. Others use the term more broadly to refer to accounts that do not posit universals, unless these are understood as classes of concrete or abstract particulars. The trope account of properties, then, as I am using the term, is not a form of nominalism. My discussion of various forms of nominalism follows Armstrong's (1978) terminology.

I develop the argument against class nominalism, but the argument could be reframed in terms of pluralities or mereological sums of objects to counter those nominalists who want to avoid commitment to sets.

Whenever I speak of 'sets of' individuals, one could substitute either 'pluralities of' or 'mereological sums of'. If one interprets 'set of' in the way that Black (1971) and Stenius (1974) suggest, the view that properties are pluralities of individuals is a form of class nominalism. The argument also applies to those who want to avoid *reducing* properties to classes of individuals, but also only countenance individuals and classes. I argue that there must be something more than concrete particulars and classes of these.

Class nominalism comes in various forms. One might conceive of properties as classes of individuals in the actual world. Such accounts of course face the problem of coextension – two distinct properties might be instantiated by the same individuals in the actual world. To resolve this problem, the classes could be expanded to include particulars in other possible worlds, as Lewis (1981) suggests. For Lewis, since individuals exist in only one world, properties can be understood as classes of individuals in all possible worlds. For those who allow individuals to exist in multiple worlds, properties can be understood as classes of ordered pairs of worlds and individuals.

My argument differs from those who link the existence of properties to the *justification* of inductive inferences. Dretske (1977), Tooley (1977, 1987), and Armstrong (1983) argue that there are universals and lawful relations between them; otherwise, we would be unable to *justify* inductive inferences or *confirm* general hypotheses. Both the premise I use and the conclusion I draw are different. Rather than assume that we can *justify* inductive generalizations, I merely assume we can *formulate* them. Moreover, I am not arguing that properties are universals nor that properties are sparse. As I discuss below, the distinction between natural and artificial properties is irrelevant to the problem I raise; and there is more than one account of properties that allows for the formulation of inductive generalizations, including trope and resemblance accounts.

My premise is that we have made inductive claims in the past. I argue that this would not be possible were there no properties over and above classes of concrete particulars. Inductive generalizations are generalizations where all the instances covered by the generalization have not yet been encountered, or at least where it is not yet known whether all the instances satisfy the consequent of the generalization. Over the course of my dog-sitting experiences, I have made the following observations:

Emerson is a dog and is friendly.
Bertie is a dog and is friendly.
Ramsey is a dog and is friendly.

I can use these observations to make generalizations. Some of these are inductive, e.g. 'All dogs are friendly'; others are not, e.g. 'All dogs I have taken care of are friendly'. When we claim, 'All dogs are friendly', it is also possible that, unbeknownst to us, we have observed the friendliness of all dogs that ever have or will exist. There are two views one might take about such situations, corresponding to two ways we might understand what an inductive claim is. We might require that inductive claims have instances not yet observed, or we might require only that they *potentially* cover unobserved instances. If the former, then my premise is that we have made inductive claims. If the latter, then my premise is that we have made inductive claims that had instances not yet observed at the time of the assertion. My premise is also not that we *think* we formulate inductive claims, but that we actually do so. Otherwise, we could only conclude that we must *believe* there are properties that are not merely classes of particulars, not that there actually are any. So my premise is:

> (P) We have formulated inductive claims that have had instances not yet observed.

For example, my niece recently asserted, 'All dogs are mean', which clearly had instances she had not yet observed.[4] Since actuality implies possibility, it follows that:

> (C1) We have the ability to formulate inductive claims that have instances not yet observed.

The remainder of the chapter will establish that we would not have this ability were there nothing over and above concrete particulars and classes of these. The argument proceeds in two phases. In section 2, I argue that (C1) entails:

> (C2) Inductive claims cannot be defined extensionally, where by this I mean in terms of particulars or sets of particulars.

I am not claiming that nominalists are actually committed to such a view. However, were such a view tenable, then class nominalism would be defensible as well. Thus, (C2) is crucial to establishing the main conclusion of the chapter, which I discuss in section 3:

> (C3) There must be properties that are not merely sets, classes, pluralities, or mereological sums of concrete particulars, whether concrete particulars in this world or those in all possible worlds.[5]

Were (C3) false, we would be unable to formulate inductive generalizations.

2.

My goal in this section is to show that inductive claims cannot be defined extensionally, i.e., in terms of particulars or sets of particulars. I argue that we can only refer to objects not yet encountered[6] if we rely on predicates that have descriptive content or refer to properties that are not merely classes of concrete particulars.

There are several ways we might try to give content to inductive generalizations in terms of particulars or sets of particulars. We might try to define the *predicates* in inductive generalizations by their extension or define the entire claim only relying on particulars or sets of particulars (or names for the particulars). Moreover, the sets of particulars might be limited to those in the actual world, or include those in other possible worlds. The problem for all these options is that inductive generalizations concern objects not yet encountered, or where it is not yet known whether all the objects mentioned in the antecedent satisfy the consequent. In either case, we would be unable to refer to such objects unless we relied on predicates that have descriptive content or refer to properties that are not merely sets of particulars. My concern is not that we cannot *know* or *have evidence about* which objects our claims are about, but with how we can *refer* to objects not yet encountered in the first place; whether we can know which objects these are is irrelevant.

Notice that the conclusion I reach is stronger than the common objection to such accounts of meaning. One might object that such extensional accounts of meaning are inadequate because they fail to capture the content of our claims, either because they individuate concepts (or propositions) in a way that is too coarse-grained or because, even if they could individuate concepts (or propositions) correctly, they would still not adequately capture their meanings. While this might be true, the claim I am defending is not simply that our claims and predicates happen to have content that cannot fully be captured in terms of particulars or sets of particulars; the claim is that at least some of our predicates (those used in inductive generalizations) and some of our claims (inductive generalizations) could not possibly be defined in terms of particulars or sets of particulars.

The simplest way to define inductive generalizations using sets of particulars is to define the *predicates* used in them by listing their extensions in the actual world. This can be done in two ways: by defining the predicate directly in terms of the set of objects or by assigning a name to each object and defining the predicate as the set of these names. Consider, first, the former. One might think the problem with this approach is that we would be unable to grasp the meaning of 'is a dog', if this is how its meaning were captured. After all, the set of dogs includes dogs that have not been observed. However, this problem cannot even arise, because it is impossible for us to define 'is a dog' as the set of dogs, if our predicates do not have descriptive content or refer to properties that are not merely sets of particulars. There would, then, be nothing that could fix the reference of 'is a dog'.

Even if we do not require that we have full access to the meanings of our predicates, there must still be something that makes it the case that we mean by 'is a dog' the set of dogs, rather than, say, the set of the three dogs I have encountered and all the future pencils. The problem is that nothing could fix this reference if we define our predicates purely in terms of particulars or sets of particulars. We cannot rely on descriptive content, since we are defining 'is a dog' purely in terms of sets of particulars; nor can we rely on causal relations, since dogs not yet encountered cannot cause our use of 'is a dog'; and we cannot intend the predicate 'is a dog' to refer to all dogs if we only rely on extensional accounts of meaning. If we intend our predicates to refer to objects not yet encountered, this intention cannot be fixed by the set that we have encountered; and, since we do not have access to the entire set of dogs, this intention cannot be specified by listing the set of all dogs. Therefore, we cannot define 'is a dog' by listing the set of dogs, since we cannot even refer to the set of dogs if the only resources we have to define predicates are particulars or sets of particulars. Moreover, this problem is not helped by expanding the set to include objects in other possible worlds. If we cannot refer to particulars not yet encountered in this world on such an account of meaning, how could we possibly refer to those in other possible worlds?

This problem also arises when all objects covered by the generalization have been encountered, but it is not known whether they all satisfy the consequent. For example, after first seeing my dog, Honey, I might have asserted, 'All dogs I've observed are friendly dogs I've observed', even though I had not yet observed Honey's behavior around people or other dogs. The claim would only cover objects already encountered, but it would still be inductive. However, while I could define 'is a dog I've observed' by listing the dogs I have observed, I could not do so for 'is a friendly dog I've observed', since it is unclear whether Honey should be included in the set of friendly dogs I've observed; and if one simply defined 'is a friendly dog I've observed' extensionally by including Honey in the set, the claim would not be inductive.

Thus, to formulate inductive generalizations, there must be some other way of referring to the set of dogs. Rather than specifying a set by listing its members, we must specify some criterion for being included in the set. There are two ways this might be done: by relying on predicates that have descriptive content or by fixing the content of our predicates by the properties to which they refer, where these properties cannot merely be sets of concrete particulars, since the very problem we are trying to solve is how we could refer to the particulars in the first place. The latter option is incompatible with class nominalism; in section 3, I argue that the former is as well. Before doing so, however, I will consider other ways of defining predicates (or sentences) in terms of sets of particulars. Note, though, that my argument does not entail that inductive claims are *about* anything more than concrete particulars or sets of these. 'All dogs are friendly' might be understood as a claim *about* concrete particulars or sets of these. However, we would be unable to make inductive claims about these objects unless some of our predicates have descriptive content or refer to properties that are not merely sets of particulars.

One might try to avoid the problem raised by using proper names for objects not yet encountered. We might try defining 'is a dog' by assigning names to all dogs, including those not yet encountered. However, this would not solve the problem. There are three main views about proper names: they are equivalent to definite descriptions, they are defined purely by their reference, or they have senses that are not descriptively capturable. The first view is unhelpful if we are trying to reduce the definitions of predicates to sets of proper names, since one cannot define predicates as sets of names and then use these predicates to give content to names; the other two options are precluded by our ability to formulate inductive generalizations.

Consider first the possibility of defining predicates in terms of names that are purely referential. We would need names for all dogs and be able to distinguish these from the sets of names used to define other predicates. In addition to not knowing how many dogs there will be, there is a more fundamental problem. I might invent names and use these, along with the names for dogs already encountered, to define 'is a dog'. However, it is unclear how these names could refer to any specific individual not yet encountered. Suppose I created the name, 'Jain', and included it in the set used to define 'is a dog'. The problem is that there is nothing that could make it the case that 'Jain' refers to (or fails to refer to) any future particular, since the name has no descriptive content. Its reference cannot be fixed causally, since an object not yet encountered cannot be the cause of my use of 'Jain'; and its reference cannot be fixed by my intentions, since I have no access to objects not yet encountered.[7]

The problem is not helped by adding names for objects in other possible worlds. If names are defined purely referentially, the reference of these names could not be fixed for objects not yet encountered, whether these referents are in the actual world or in another possible world. In fact, the problem is particularly pernicious for counterpart theories. Unless we rely on descriptions or properties that are not classes of particulars, we would have no way of fixing the reference of names of counterparts of particulars already encountered. Therefore, predicates cannot be defined by sets of names that are themselves defined purely referentially.

Nor could we rely on proper names that have senses that are not descriptively capturable. This would simply make it impossible for us to come up with names for such objects, since each sense will be peculiar to each particular. Moreover, in order for our claim, 'All dogs are friendly', to be *about* dogs, we would need senses for all the dogs not yet encountered and be able to distinguish these from senses for particulars that are not dogs. Again, this problem cannot be solved by arguing that we need not have complete access to the meanings of our terms. Even if this is so, there would still need to be something that makes it the case that our claim is about dogs. However, this would be impossible if all we had available to fix the meaning of 'is a dog' were senses for dogs already encountered. Therefore, if senses for particulars were all we had to fix the meanings of our predicates, it would be impossible for us to make assertions about objects not yet encountered.

Rather than defining the *predicates* used in inductive claims extensionally, we might define the *entire claim* by relying on particulars or sets of particulars. Often, this is done in terms of possible worlds, for example by defining the proposition expressed by a generalization as a set of possible worlds (i.e. those where the proposition is true). However, this faces the same problem discussed above, except now the question is how we can refer to possible worlds without relying on predicates either with descriptive content or that refer to properties that are not merely sets of particulars. If no individual exists in more than one possible world, it seems we could pick out possible worlds by the individuals they contain, rather than relying on such predicates. However, for reasons already discussed, without such predicates, we would be unable to refer to individuals in other possible worlds. Alternatively, if an individual can exist in more than one world, we could not pick out a possible world by specifying which individuals they contain; and, while we might be able to refer to the actual world, it is unclear how we could refer to other possible worlds without predicates that either refer to properties that are not merely sets or have descriptive content.

3.

We seem to be faced with two options: predicates used in inductive generalizations must either have descriptive content or refer to properties that are not merely sets. The latter option is clearly incompatible with class nominalism. As I will now argue, the former is as well.

In order to formulate inductive generalizations, it must be *possible* for such claims to be *about* objects not yet encountered or about objects already encountered, but where it is not known whether all objects satisfying the antecedent also satisfy the consequent. Thus, predicates used in inductive claims must have some criteria of application in virtue of which they are able to apply to such objects. We might not require that the predicates actually refer to such objects. However, it must be possible for them to do so. I begin with a discussion of what this entails for claims about objects not yet encountered. I then consider cases where we have encountered the objects, but not yet determined whether the consequent holds of them.

There are several reasons predicates might fail to refer to objects not yet encountered. First, as argued, if predicates were defined extensionally, it would be impossible to refer to objects not yet encountered. Alternatively, the *world* might be such that our predicates fail to refer to objects not yet encountered. There are (at least) two reasons this might occur. The first does not preclude our ability to formulate inductive generalizations; the second does. First, we might assert an inductive claim, but there happen to be no objects not yet encountered that satisfy the predicates used in the claim. The predicates might either have no referents or have referents, but all those referents have been observed. Scientists made claims using the predicate 'is phlogiston', even though there is no phlogiston. One can also imagine someone on Mauritius in the late 1600s viewing the last living dodo (not realizing it is the last) asserting the claim, 'All living dodos will be found on Mauritius'. Such a claim, while applying to the last living dodo, would apply only to objects already observed. Whether one counts these claims as inductive (see section 1), they do not preclude our ability to make inductive generalizations. We can still formulate inductive claims about other objects. We can even formulate inductive claims that have instances not yet observed using 'is phlogiston' and 'is a dodo'. Before dodos were extinct, people were able to make claims applying to future dodos. Moreover, scientists were able to assert, 'All flammable materials contain phlogiston'. Such cases could preclude our ability to formulate inductive claims (assuming one even requires that inductive claims have instances not yet encountered) only if it were impossible for us to formulate predicates that apply to *any* objects whatsoever (i.e., if all our predicates were like 'is phlogiston'). Not only is this clearly not so, but if it were, then our problems would be more serious – we would be unable to make *any* true claims (inductive or otherwise) using such predicates.

The problem I am concerned with arises if there were no properties over and above sets of concrete particulars – if class nominalism were true. As noted, in order for claims to be inductive, it must be possible for the predicates used in the claim to refer to objects not yet encountered. Therefore, there must be some criteria of application in virtue of which the predicates can apply to such objects. If predicates used in inductive claims could be defined in terms of sets of particulars, then whether an object satisfies a predicate would simply be a matter of determining whether the object (or name) is contained in the set. However, as argued, predicates used in inductive claims cannot be so defined. Instead, such predicates must either have descriptive content or refer to properties that are not merely sets of particulars. We can ignore the latter option, since the question is whether formulating inductive claims would be possible were class nominalism true.

Consider, then, the consequences if our predicates were defined at least partly descriptively, and there were no properties over and above sets of particulars. I say at least partly descriptively, because we might define the predicates partly by listing the objects already observed. Consider the claim, 'All dogs are friendly'. I might define 'is a dog' partly by listing objects already encountered and add descriptive content to cover other cases. The description might simply be 'and all objects that resemble these'. The problem is that, were there no properties over and above sets of concrete particulars, it would be impossible for an object not yet encountered to satisfy the predicate. There would be no universals, tropes, or even resemblance relations between the concrete particulars. There would be nothing *about* the objects not yet encountered that would enable them to satisfy predicates used in inductive claims. Moreover, it does not help to claim that particulars are thick particulars, as some nominalists[8] have argued, since according to class nominalism this thickness would be cashed out in terms of membership in a class.[9]

One might say that the objects satisfy the predicate 'is a dog' because they each have the property of being a dog. But if properties are merely sets, this amounts to saying they are members of a particular set. The question is what makes it the case that one particular set, as opposed to some other set, corresponds to 'is a dog'. The predicate is not defined by the set, and there is nothing about the objects in one set that makes it the case that they, as opposed to some other set of objects, satisfy the descriptive content of the predicate.

The same applies to claims about objects already encountered, but where it is unknown whether the consequent is satisfied by all objects satisfying the antecedent. Recall the example of Honey. Before observing Honey's behavior, I might have asserted: 'All dogs I've observed are friendly dogs I've observed'. As argued in the previous section, while I could define 'is a dog I've observed' by listing the dogs I have observed, I could not do so for 'is a friendly dog I've observed'. Consequently, 'is a friendly dog I've observed' must have descriptive content (assuming it is not defined by a property to which it refers), and there must be something about Honey that makes it the case that she does or does not satisfy this predicate, where this cannot simply be a matter of belonging to a set.

My concern is not that we need an analysis of the class membership relation or an analysis of what makes a class natural, both of which Armstrong (1978) discusses. We might take both as primitive, as Lewis (1981) suggests. The problem persists. It may be a primitive fact that a particular belongs to some class or that some classes are natural. My concern is how we could ever refer to any one of these classes, if there were no properties over and above classes of particulars. The question is what makes it the case that one set, as opposed to some other set, corresponds to 'is a dog'. There is nothing about the objects in one set that makes it the case that they, as opposed to some other set of objects, satisfy the descriptive content of the predicate. Lewis (1981) argues that a primitive distinction between natural and artificial classes allows classes to do the work of other accounts of properties. However, the argument of this chapter shows that we cannot even pick out a class of particulars that includes objects not yet encountered (natural or not) unless there are properties that are not merely sets of particulars. Designating some classes as natural does nothing to resolve this problem. Thus, much of the work Lewis wants properties to do could not be accomplished were class nominalism true.

It also does not help to include objects in other possible worlds, as Lewis (1981) argues we might. We need predicates to apply to objects not yet observed in the *actual* world, since this is what inductive generalizations are about. Having a set that includes objects in other possible worlds does nothing to fix whether an object in the actual world satisfies a predicate defined descriptively. Whether an object satisfies or fails to satisfy such predicates cannot be a matter of their being contained in a set, whether the set includes objects in the actual world or also those in other possible worlds. There must be something about objects in the *actual* world that makes it the case that they satisfy or fail to satisfy such predicates.

The argument does not entail that all (or any) predicates used in inductive generalizations refer to natural properties or to properties at all. As Bigelow (1981) argues, what explains *why* a term signifies something does not entail that this is *what* the term signifies. We might say our predicates refer to sets of particulars, and yet claim that they do so, at least in the case of inductive generalizations, only because the particulars have properties that are not just classes of particulars. In fact, we need not even understand our predicates to be referring expressions at all (see the discussion below about ostrich nominalism). As I framed the argument above, our predicates *apply* to objects not yet encountered, but this does not entail that our predicates designate these objects in the way proper names would.

Moreover, we can formulate inductive generalizations that use non-natural predicates, such as 'is grue', or that use non-referring predicates, such as 'is phlogiston'. We can formulate inductive generalizations that have no referent and that pick out non-natural classes. This is why Goodman's new riddle of induction has force. Additionally, as Davidson (1966) argues, we can even formulate *true* inductive generalizations using grue-like predicates, as in 'All emeroses are gred'. This does not entail there is a property corresponding to 'is an emerose'. Nevertheless, in order for 'All emeroses are gred' to be a claim about objects not yet encountered, there must be something about the objects, some property that the objects do have (e.g. being an emerald), such that 'is an emerose' is able to refer to these objects. Were there no properties, or were properties merely classes, this would be impossible.

One might object that predicates used in inductive generalizations might not always have clear criteria of application. In other words, there could be cases where it is unclear whether a particular object satisfies the predicate. In such cases, the generalization itself might not have a determinate truth value. My argument, though, does not depend on inductive generalizations having determinate truth values. Suppose that all predicates have cases for which it is unclear whether they apply. (As I discuss below, this might arise for resemblance nominalism.) There might, then, be no inductive generalizations that have determinate truth values. However, for a generalization to be inductive, it must potentially be about *some* objects not yet observed. Therefore, there must be *some* objects not yet observed for which the application is determinate, and this requires that there be something over and above concrete particulars and classes of these.

I have established that class nominalism is incompatible with our ability to formulate inductive generalizations. The argument applies equally to those who want to avoid a commitment to sets, but allow for pluralities or mereological sums of concrete particulars. There still must be something about the objects in virtue of which the predicates used in inductive claims are able to apply to objects not yet encountered. If there were no properties over and above mereological sums or pluralities of particulars, there would be nothing about the objects we could specify prior to encountering the objects that would allow us to make claims about them. Likewise, both predicate and concept nominalism are incompatible with our ability to formulate inductive generalizations, since, like class nominalism, there would be no features of objects in virtue of which objects could satisfy predicates used in inductive generalizations.

The form of nominalism defended most notably by Quine (1948) and Devitt (1980), dubbed 'ostrich nominalism' by Armstrong, is also problematic. According to this view, we can understand concrete particulars to be thick, and can assert, 'Ramsey is a dog', and even, 'There is a property, doghood, that Ramsey and Bertie share', without being committed to anything more than particulars. We can avoid the apparent commitment to doghood in our assertion, 'There is a property, doghood, that Ramsey and Bertie have in common', by rephrasing it as, 'Ramsey is a dog, and Bertie is a dog'. 'Ramsey is a dog' is true iff 'Ramsey' designates Ramsey and 'is a dog' applies to Ramsey. Thus, we are only committed to the existence of Ramsey and the predicate 'is a dog'. Whether this works as a response to the one-over-many problem (its intended target), it does not help with the problem raised here. This sort of analysis is impossible with inductive generalizations. When uttering inductive generalizations, we have no names for many of the objects to which the predicates apply. The concern is not with the ontological commitments of claims of the form 'Fa', since we are unable to make claims of this form about objects not yet encountered. Moreover, it does not help to argue that, upon encountering objects, we can assign a name and at that point offer this analysis. The point is that we are able *now* to make claims about these objects, though we have not encountered them and even if we *never* encounter them. Thus, this analysis is unhelpful in such cases.

Most other accounts of properties are acceptable. I discuss possible ways other accounts might avoid the problem raised. However, this does not entail that these are the only or best ways of doing so. If there are universals, these universals provide criteria by which predicates refer (or fail to refer) to particulars not yet encountered. For example, we might define the predicates used in inductive claims referentially by the universals to which they refer.[10] Since universals are wholly present in each object that instantiates it, by specifying which universal (or universals) we are talking about, we are able to make claims about all objects that instantiate that universal, whether we have encountered those objects or not.

Trope accounts are also adequate. We might define a predicate partly by tropes already observed, and simply add the phrase 'and all other tropes that exactly resemble these'. Particulars not yet observed would then satisfy the predicate if they exactly resemble the tropes already encountered. Likewise, the resemblance nominalist can define predicates by taking particulars already encountered as exemplars and adding the phrase 'and all those objects resembling the exemplars'. This resemblance relation differs according to different resemblance accounts. I might take Ramsey, Bertie, and Emerson as exemplars for our definition of 'is a dog'. According to Price's (1953) account, when I claim that all dogs are friendly, I am making a claim about all objects resembling Ramsey, Bertie, and Emerson at least as much as Ramsey, Bertie, and Emerson resemble each other (see also Carnap, 1967; Hausman, 1979; Lewis, 1981;[11] and Rodriguez-Pereyra, 2002). One question arising for all resemblance accounts is whether the objects encountered are adequate to specify a unique set. However, this would not preclude our ability to make inductive claims. Imagine that the phrase, 'all objects resembling Ramsey, Bertie, and Emerson at least as much as Ramsey, Bertie, and Emerson resemble each other', does not pick out a unique set of particulars. So long as there is at least one object not yet encountered overlapping all these sets, our claim would be about this object, even if it was ambiguous about other objects. The claim would then be an inductive claim about at least one object not yet encountered. Whether this occurs is an empirical question. If the world is uncooperative, resemblance nominalism might preclude us from making inductive claims that cover cases not yet observed. If we could show that we have made an inductive claim that covers cases not yet encountered at the time the claim was made, and if we could establish that no objects overlap all sets specified by whatever resemblance we choose, this would show that resemblance nominalism is untenable. Such a task, however, goes beyond the bounds of this chapter.

4.

While this argument shows that most versions of nominalism are inadequate, it does not establish which account of properties is correct. For the issue at hand, this need not be decided. What I have shown is that formulating inductive generalizations would be impossible unless objects have universals, tropes, or stand in the appropriate sort of resemblance relations.

There is another general lesson to be drawn from the foregoing arguments about how we ought to decide metaphysical or ontological questions. Quine and others argue that ontological or metaphysical questions about what there is ought to be settled at the end of science.[12] One must first establish which theories are true; only then can one decide what ontological or metaphysical commitments to accept by determining what is required for the theories to be true. However, the problem discussed in this chapter would not arise at the end of science, anymore than it would for god-like creatures who have access to all objects that have and will exist. Such creatures do not formulate inductive generalizations. The same would be true for us at the end of science, assuming we were no longer making claims about objects not yet encountered. We are not god-like creatures, though, and we do not currently have access to all objects potentially covered by our generalizations. This is why we formulate inductive generalizations. Luckily we live in a world where this is possible – a world where there are properties. What this shows is that there are metaphysical requirements for the activities we engage in during the development of science. It is true that at the end of science we could reconstruct the metaphysical assumptions differently. However, to do so would overlook what allows for the development of scientific knowledge in the first place.

Acknowledgements

I would like to thank Jim Bogen, Jacob Busch, Sue Dwyer, Philip Kitcher, P. D. Magnus, Paul Pietroski, Jim Thomas, and Steve Yalowitz for comments on earlier drafts of this chapter.

Notes

1. My discussion will focus on properties, but the argument applies equally to relations.
2. The difference between sets and classes is irrelevant to the issues discussed here, so I will use these terms interchangeably.
3. Moreland (1989) calls this view 'extreme nominalism'.
4. It makes no difference whether the instances confirm or falsify the generalization.
5. To avoid quantifying over properties or to avoid understanding these various accounts as *constitutive* of properties, one could rephrase my conclusion as: (C3*) Objects must instantiate universals or tropes, or stand in the appropriate resemblance relations. I argue below that all three of these accounts are compatible with our ability to formulate inductive claims.
6. I will use the terms 'observe' and 'encounter' interchangeably, though I do not mean to imply that observation is required for fixing reference.
7. This is similar to Lewis's argument (1986b: 158) against ersatz accounts of possible worlds for lacking the descriptive power to discriminate all possibilities.
8. I discuss below why this also will not help other nominalists, such as Devitt (1980). Defenders of such a view, which Armstrong has classified as ostrich nominalism, are typically not trying to give an account of what it is for an object to have a property or to satisfy a predicate.
9. The class nominalist could in theory be a class nominalist only about universals, and then give a different account (or none at all) of what it is for a particular to be thick. Whether the latter helps with the problem I have raised depends on what account one gives. I address this below by considering whether other accounts of properties can resolve the problem I raise.
10. This does not require that each predicate refers to a single universal (as with 'is grue').
11. Lewis (1981) argues that class nominalism and his version of resemblance nominalism are interchangeable. If my argument is correct, his version either would be unable to solve the problem raised or is not equivalent to class nominalism.
12. *Our* current metaphysical commitments (what we must believe to exist or be true given our current assertions) can be decided now, according to Quine. However, what is actually true of the world is determined by what sentences are true, not what sentences we believe.

References

Armstrong, D. (1978) *Nominalism and Realism: Universals and Scientific Realism Volume I* (Cambridge: Cambridge University Press).

Armstrong, D. (1983) *What is a Law of Nature?* (Cambridge: Cambridge University Press).

Bigelow, J. (1981) 'Semantic Nominalism', *Australasian Journal of Philosophy*, 59: 403–421.

Black, M. (1971) 'The Elusiveness of Sets', *Review of Metaphysics*, 24: 614–636.

Carnap, R. (1967) *The Logical Structure of the World* (Los Angeles: University of California Press).

Davidson, D. (1966) 'Emeroses by Other Names', *Journal of Philosophy*, 63: 778–780.

Devitt, M. (1980) ' "Ostrich Nominalism" or "Mirage Realism"?', *Pacific Philosophical Quarterly*, 61: 433–439.

Dretske, F. (1977) 'Laws of Nature', *Philosophy of Science* 44: 246–268.

Hausman, A. (1979) 'Goodman's Perfect Communities', *Synthese*, 41: 185–237.

Lewis, D. (1981) 'New Work for a Theory of Universals', *Australasian Journal of Philosophy*, 61: 343–377.

Lewis, D. (1986a) 'Against Structural Universals', *Australasian Journal of Philosophy*, 64: 25–46.

Lewis, D. (1986b) *On the Plurality of Worlds* (Oxford: Blackwell).

Moreland, J. P. (1989) 'Keith Campbell and the Trope View of Predication', *Australasian Journal of Philosophy*, 67: 379–393.

Price, H. H. (1953) *Thinking and Experience* (Cambridge, MA: Harvard University Press).

Quine, W. V. (1948) 'On What There Is', *Review of Metaphysics*, reprinted in his 1953 *From a Logical Point of View* (Cambridge, MA: Harvard University Press).

Rodriguez-Pereyra, G. (2002) *Resemblance Nominalism: a Solution to the Problem of Universals* (Oxford: Clarendon Press).

Stenius, Eric (1974) 'Sets', *Synthese* 27: 161–188.

Tooley, M. (1977) 'The Nature of Laws', *Canadian Journal of Philosophy*, 7(4): 667–698.

Tooley, M. (1987) *Causation* (Oxford: Oxford University Press).

5
Models and Scientific Representations

Otávio Bueno

1. Introduction

Scientific theories and the models that characterize them are often taken to be representational devices. But how exactly do they represent? It is unclear whether there is a unique form of representation in scientific practice, given that so many different means of representation are used. Although models are central in scientific representation, they are certainly not the only resource available: outputs of various instruments, micrographs, templates, diagrams, and a variety of other items are also used. And given the plurality of these means of representation, different representation relations are then established among such means and the corresponding targets; for example, different structural relations between these items are often invoked (such as various partial morphisms).

This doesn't mean, however, that nothing general can be said about scientific representations. In this chapter, I offer a framework in terms of which the multiplicity of representations can be accommodated, but which still preserves the differences among the various means of representation and the many representation relations. After discussing the details of the framework, in terms of partial structures and partial relations, I'll illustrate how the framework can be put to work, by examining some examples from microscopy that illuminate important aspects of scientific practice.

2. Scientific representation: a framework

The first step consists in developing a framework to examine the notion of representation. Broadly understood, representation involves three central components: (i) Representation is an *intentional act* relating two objects: the *source* (a certain object or event that is used to represent something) and the *target* (a given object or event that is represented). (ii) Representation also involves a *coding convention* – a certain group of rules that specify what counts as the right *similarity relation* between the source and the target. (iii) And, finally, a *mechanism of representation* is also involved. This mechanism establishes how the target is represented: e.g. the target is represented *as* such and such *in virtue of* suitable features of the source (for example, in virtue of mappings that have suitable properties).

Consider, for instance, the case in which a given model (e.g. a particular graph) is used to represent certain phenomena (say, the average growth of the US economy). In this case, the source is the graph; the target is the average growth of the American economy; the coding convention indicates that an increase (or a decrease) in the slope of the curve in the graph corresponds to an increase (or a decrease) in the average growth of the economy. Suppose now that a continuous line characterizes the graph under consideration. The US economy's growth, in this case, is represented *as* continuous *in virtue of* the continuity of the line in the graph.

Jill Sigman and Bas van Fraassen have elegantly presented these features of representation by offering a basic format for representation:

> Representation of an object involves producing *another object* which is *intentionally* related to the first by a certain *coding convention* which determines what counts as *similar* in the *right way*. (1993: 74; italics added)

As the passage makes clear, representation is understood as an inherently intentional act that relies crucially on coding conventions to specify in which respects the target and the source are similar. This proposal offers a broad account of representation, independently of the context in which the notion happens to be used.

Two additional features of representation need to be emphasized. Although significant for representation in general, these features are especially relevant for scientific representation. (a) First, it's important to emphasize that representation is always *perspectival*; it emerges from, and presupposes, a given perspective. That is, a representation is always established from a certain point of view – highlighting a certain standpoint. For instance, when we represent a physical system as a Newtonian system, we emphasize certain components of the system; we are adopting a certain perspective in terms of which the system is represented. The representation is informed by the conceptual resources of Newtonian physics, being in general silent about those components that are not identified as relevant from the vantage point of the Newtonian framework. The representation is likely to emphasize features such as the mass, velocity, and acceleration of the objects under study, not saying much (if anything at all) about their color or smell.

However, we need more than the perspective from which the system is represented. We also need to locate ourselves in the model that is used to represent the system. In other words, representation depends on an *indexical feature*. Just as to use a map, we need to identify ourselves with respect to the map, to use a model we also need to identify ourselves with respect to the model. That is, we need to locate ourselves and the situation we are currently experiencing with respect to the resources of the model. As van Fraassen points out:

> To use a theory or model, to base predictions on it, we have to locate ourselves with respect to it. We have to be able to say, for example, that the phenomenon *we are presently witnessing* is classified in the theory as oxidizing, or as phlogiston escape, or the like. We have to locate our situation in the theory's logical space, in a way that is similar to our 'We are here' with respect to a map. (van Fraassen, 2006: 31)

In other words, in order to extract the information that the map displays, we need to specify our location with respect to the map. This step is crucial: it's only relative to that location that the information provided in the map can be used. In a similar way, to be able to use a theory or a model, we need to locate ourselves with respect to such items. By identifying the phenomenon that we are currently experiencing as something that the theory classifies in a certain way, we are then able to use the information provided by the theory to describe the phenomenon. This is an indexical component – where we are with respect to the model – and this component highlights a pragmatic point regarding the use of the model in question.

(b) Second, it's also important to emphasize that representation typically involves *interpretation* (Sigman and van Fraassen, 1993). To represent a certain phenomenon *as* such and such, we need to interpret the phenomenon in a certain way – we need to understand and conceptualize it suitably. Given the perspectival nature of scientific representation, interpretation will always be involved. After all, the ability to make sense of the phenomena in terms of the conceptual resources of a theory or model is required. It's in light of the theory that the representation is ultimately established. But in order to do that, the phenomenon needs to be interpreted according to the theory. When Galileo refused to represent the fall of bodies in purely qualitative, Aristotelian terms, and adopted instead a mathematical description of their fall, he changed the interpretation of the phenomena. By shifting the interpretation, Galileo also modified their representation. Falling bodies were then represented as satisfying certain mathematical regularities – in terms of a very different framework from the one adopted by the Aristotelians.

So far, I have described representation in general, and scientific representation in particular, as a certain product: an entity that emerges from taking intentionally a certain object as standing for another in a certain way. This way of speaking, however, can be misleading. It invites us to reify the notion of representation. It invites us to think of representation as a certain object. But reification, particularly of abstract entities, is something that several of us – empiricists, nominalists, fictionalists of various forms – work very hard at avoiding. But can that be avoided here?

I think it can. Consider representation as a derivative notion: representation is what emerges from a certain activity – the activity of representing. What is involved in that activity? The activity involves intentional acts, acts of taking something to stand for something else, given (i) a certain way of conceptualizing the objects in question (a particular interpretation), and (ii) certain coding conventions (that determine the respects in which the objects in question are similar).

If we read carefully the quotation above from Sigman and van Fraassen (1993) on the format for representation, we see that it already indicates that what matters is the activity of representing. Note how the emphasis lies in the activity of *producing* something intentionally: 'Representation of an object', Sigman and van Fraassen note, 'involves *producing another object* which is intentionally related to the first by a certain coding convention' (1993: 74; italics added). What is central here is not an object – the representation – but a certain intentional activity of taking something to stand for something else – the representing. Ultimately, the latter gets the work done rather than the former. And the same emphasis on representing should be found when we consider representation in scientific practice.[1]

3. Scientific representation: mapping partial structures

What should be said then about the mechanism of representation in the case of representation in science? What sorts of relations should be established between the target and the source for scientific representation to be established? The answer, of course, is that there are many such relations. But is there something more substantive, but still general that could be said? I think there is. What follows is a structural account of a family of these relations that offers a way of accommodating some of the central mechanisms of representation involved in scientific practice. Science is, of course, such a complex phenomenon that it's unlikely that a single account could be offered to cover every aspect of scientific practice. But some generality is still possible without being entirely comprehensive.

Models are central vehicles of scientific representation. And the striking feature of these models is often what they leave out, which aspects of the phenomena that the models are taken to represent are, in the end, left untouched. It's often only by ignoring various features of the phenomena under study that models become tractable. If too many details are included in the model, very quickly the latter can no longer function as a model – a device that helps us identify and study central features of the phenomena in which we are interested. So, the partiality of the information that is included in a model is crucial (see da Costa and French, 2003).

This partiality can be accommodated formally in terms of the partial structures approach (see Bueno, 1997; Bueno et al., 2002; da Costa and French, 2003). This approach relies on three main concepts: partial relation, partial structure, and quasi-truth. One of the main motivations for introducing this proposal derives from the need for supplying a formal framework in which the openness and incompleteness of the information that is dealt with in scientific practice can be accommodated. This is accomplished, first, by extending the usual notion of structure, in order to accommodate the partialness of information we have about a certain domain (introducing then the notion of a partial structure). Second, the Tarskian characterization of the concept of truth is generalized for partial contexts, which then leads to the introduction of the corresponding concept of quasi-truth.

The first step, then, to characterize partial structures is to formulate a suitable concept of a partial relation. In order to investigate a certain domain of knowledge Δ (say, the physics of particles), researchers formulate a conceptual framework that helps them systematize and interpret the information they obtain about Δ. This domain can be represented by a set D of objects (which includes *real* objects, such as configurations in a Wilson chamber and spectral lines, and *ideal* objects, such as quarks). D is studied by the examination of the relations that hold among its elements. However, it often happens that, given a relation R defined over D, we do not know whether all objects of D (or n-tuples thereof) are related by R, or we need to ignore some of the relations that are known to hold among objects of D, in order to study other relations about that domain in a tractable way. This is part of the incompleteness and partiality of our information about Δ, and is formally accommodated by the concept of a partial relation. The latter can be characterized as follows. Let D be a non-empty set. An n-place *partial relation* R over D is a triple $\langle R_1, R_2, R_3 \rangle$, where R_1, R_2, and R_3 are mutually disjoint sets, with $R_1 \cup R_2 \cup R_3 = D^n$, and such that: R_1 is the set of n-tuples that (we know that) belong to R, R_2 is the set of n-tuples that (we know that) do not belong to R, and R_3 is the set of n-tuples for which it is not known (or, for reasons of simplification, it is ignored that it is known) whether they belong or not to R. (Notice that if R_3 is empty, R is a usual n-place relation that can be identified with R_1.)

But in order to accommodate the information about the domain under study, a concept of structure is needed. The following characterization, spelled out in terms of partial relations and based on the standard concept of structure, offers a concept that is broad enough to accommodate the partiality usually found in scientific practice. A *partial structure* A is an ordered pair $\langle D, R_i \rangle_{i \in I}$, where D is a non-empty set, and $(R_i)_{i \in I}$ is a family of partial relations defined over D.[2]

We have now defined two of the three basic concepts of the partial structures approach. In order to spell out the last one (quasi-truth), we will need an auxiliary notion. The idea here is to use the resources supplied by Tarski's definition of truth. But since the latter is only defined for full structures, we have to introduce an intermediary notion of structure to link partial to full structures. This is the first role of those structures that extend a partial structure A into a full, total structure (which are called A-normal structures). Their second role is model-theoretic, namely to put forward an interpretation of a given language and to characterize semantic notions. Let $A = \langle D, R_i \rangle_{i \in I}$ be a partial structure. We say that the structure $B = \langle D', R'_i \rangle_{i \in I}$ is an *A-normal structure* if (i) $D = D'$, (ii) every constant of the language in question is interpreted by the same object both in A and in B, and (iii) R'_i extends the corresponding relation R_i (in the sense that, each R'_i, supposed of arityn, is defined for all n-tuples of elements of D'). Note that, although each R'_i is *defined* for all n-tuples over D', it holds for some of them (the R'_{i1}-component of R'_i), and it doesn't hold for others (the R'_{i2}-component).

As a result, given a partial structure A, there are several A-normal structures. Suppose that, for a given n-place partial relation R_i, we don't know whether $R_i a_1 \ldots a_n$ holds or not. One of the ways of extending R_i into a full R'_i relation is to look for information to establish that it *does* hold;

another way is to look for contrary information. Both are prima facie possible ways of extending the partiality of R_i. But the same indeterminacy may be found with other objects of the domain, distinct from $a_1, \ldots, a_n$ (for instance, does $R_i b_1 \ldots b_n$ hold?), and with other relations distinct from R_i (for example, is $R_j b_1 \ldots b_n$ the case, with $j \neq i$?). In this sense, there are *too many* possible extensions of the partial relations that constitute A. Therefore, we need to provide constraints to restrict the acceptable extensions of A.

In order to do that, we need first to formulate a further auxiliary notion (see Mikenberg et al., 1986). A *pragmatic structure* is a partial structure to which a third component has been added: a set of accepted sentences P, which represents the accepted information about the structure's domain. (Depending on the interpretation of science that is adopted, different kinds of sentences are to be introduced in P: realists will typically include laws and theories, whereas empiricists will add mainly certain regularities and observational statements about the domain in question.) A *pragmatic structure* is then a triple $A = \langle D, R_i, P\rangle_{i\in I}$, where D is a non-empty set, $(R_i)_{i\in I}$ is a family of partial relations defined over D, and P is a set of accepted sentences. The idea is that P introduces constraints on the ways that a partial structure can be extended (the sentences of P hold in the A-normal extensions of the partial structure A).

Our problem is: given a *pragmatic* structure A, what are the necessary and sufficient conditions for the existence of A-normal structures? Here is one of these conditions (Mikenberg et al., 1986). Let $A = \langle D, R_i, P\rangle_{i\in I}$ be a pragmatic structure. For each partial relation R_i, we construct a set M_i of atomic sentences and negations of atomic sentences, such that the former correspond to the n-tuples that satisfy R_i, and the latter to those n-tuples that do not satisfy R_i. Let M be $\cup_{i\in I} M_i$. Therefore, a pragmatic structure A admits an A-normal structure if and only if the set $M \cup P$ is *consistent*.

Assuming that such conditions are met, we can now formulate the concept of quasi-truth. A sentence α is *quasi-true* in a pragmatic structure $A = \langle D, R_i, P\rangle_{i\in I}$ if there is an A-normal structure $B = \langle D', R'_i\rangle_{i\in I}$ such that α is true in B (in the Tarskian sense). If α is not quasi-true in A, we say that α is *quasi-false* in A. Moreover, we say that a sentence α is *quasi-true* if there is a pragmatic structure A and a corresponding A-normal structure B such that α is true in B (according to Tarski's account). Otherwise, α is *quasi-false*.

The idea, intuitively speaking, is that a quasi-true sentence α does not describe, in a thorough way, the whole domain that it is concerned with, but only an aspect of it: the one that is delimited by the relevant partial structure A. After all, there are several different ways in which A can be extended to a full structure, and in some of these extensions α may not be true. Thus, the concept of quasi-truth is strictly weaker than truth: although every true sentence is (trivially) quasi-true, a quasi-true sentence may not be true (since it may well be false in certain extensions of A).

It may be argued that because quasi-truth has been defined in terms of full structures and the standard notion of truth, there is no gain with its introduction. But there are several reasons why this is *not* the case. First, as was just seen, despite the use of full structures, quasi-truth is weaker than truth: a sentence that is quasi-true in a particular domain – that is, with respect to a given partial structure A – may not be true if considered in an extended domain. Thus, we have here a sort of underdetermination – involving distinct ways of extending the same partial structure – that makes the concept of quasi-truth especially appropriate for empiricists. Second, one of the points of introducing the concept of quasi-truth, as da Costa and French (2003) have argued in detail, is that in terms of this notion, a formal framework can be advanced to accommodate the openness and partialness that is typically found in science. Bluntly put, the actual information at our disposal about a certain domain is captured by a *partial* (but not full) structure A. Full, A-normal structures represent ways of extending the actual information which are possible according to A. In this respect, the use of full structures is a semantic expedient of the framework (in order to provide a definition of quasi-truth), but no epistemic import is assigned to them. Third, full structures can be ultimately dispensed with in the formulation of quasi-truth, since the latter can be characterized in a different way, but still preserving all its features, independently of the standard Tarskian type account of truth (Bueno and de Souza, 1996). This provides, of course, the strongest argument for the dispensability of full structures (as well as of the Tarskian account) vis-à-vis quasi-truth. Therefore, full, A-normal structures are entirely inessential; their use here is only a convenient device.

To illustrate the use of quasi-truth, let us consider an example. As is well known, Newtonian mechanics is appropriate to explain the behavior of bodies under certain conditions (say, bodies that, roughly speaking, have a low velocity with respect to the speed of light, that are not subject to strong gravitational fields etc.). But with the formulation of special relativity, we know that if these conditions are not satisfied, Newtonian mechanics is false. In this sense, these conditions specify a family of partial relations, which delimit the context in which Newtonian theory holds. Although Newtonian mechanics is not true (and we know under what conditions it is false), it is *quasi-true*; that is, it is true in a given context, determined by a pragmatic structure and a corresponding *A*-normal one (see da Costa and French, 2003).

But what is the *relationship* between the various partial structures articulated in a given domain? Since we are dealing with partial structures, a second level of partiality emerges: we can only establish *partial* relationships between the (partial) structures at our disposal. This means that the usual requirement of introducing an isomorphism between theoretical and empirical structures (see van Fraassen, 1980: 64) can hardly be met. After all, researchers typically lack full information about the domains they study. Thus, relations weaker than full isomorphism (and full homomorphism) need to be introduced (Bueno, 1997; French and Ladyman, 1997, 1999).

In terms of the partial structures approach, however, appropriate characterizations of *partial isomorphism* and *partial homomorphism* can be offered (see Bueno, 1997; Bueno et al., 2002; French and Ladyman, 1999). And given that these notions are more open-ended than the standard ones, they accommodate better the partiality of structures found in scientific practice.

Let $S = \langle D, R_i \rangle_{i \in I}$ and $S' = \langle D', R'_i \rangle_{i \in I}$ be partial structures. So, each R_i is a partial relation of the form $\langle R_1, R_2, R_3 \rangle$, and each R'_i a partial relation of the form $\langle R'_1, R'_2, R'_3 \rangle$.[3]

We say that a partial function[4] $f\colon D \rightarrow D'$ is a *partial isomorphism* between S and S' if (i) f is bijective, and (ii) for every x and $y \in D$, $R_1xy \leftrightarrow R'_1f(x)f(y)$ and $R_2xy \leftrightarrow R'_2f(x)f(y)$. So, when R_3 and R'_3 are empty (that is, when we are considering total structures), we have the standard notion of isomorphism.

Moreover, we say that a partial function $f\colon D \rightarrow D'$ is a *partial homomorphism* from S to S' if for every x and every y in D, $R_1xy \rightarrow R'_1f(x)f(y)$ and $R_2xy \rightarrow R'_2f(x)f(y)$. Again, if R_3 and R'_3 are empty, we obtain the standard notion of homomorphism as a particular case.

There are two crucial differences between partial isomorphism and partial homomorphism. First, a partial homomorphism does not require that the domains D and D' of the partial structures under study have the same cardinality. Second, a partial homomorphism does not map the relation R'_i into a corresponding relation R_i. Clearly a partial homomorphism establishes a much less strict relationship between partial structures.

Partial isomorphism and partial homomorphism offer mappings among partial structures that are less tight than their corresponding full counterparts – isomorphism and homomorphism. Partial mappings, as transformations that connect different partial models that may be used in scientific practice, allow for the transferring of information from one domain into another – even when the information in question is incomplete. After all, if a sentence is quasi-true in a given partial structure *S*, it will also be quasi-true in any partial structure that is partially isomorphic to *S* (see Bueno, 2000).

As a result, partial mappings can be used as mechanisms of representation in scientific practice. It is in virtue of the fact that certain models of a given phenomenon share some of the structure of the latter – in the sense that there is a partial mapping between the two – that these models can be used to represent the relevant features of the phenomenon under study. Of course, which features are relevant is a pragmatic matter, largely dependent on the context under consideration.

It is also possible to accommodate the significance of models in reasoning about the phenomena, even when only partial information is available about the objects under study. The partial information is encoded in a partial structure that represents selected aspects of the phenomena. As noted above, representation is always made from a particular perspective; it is, thus, selective. In other words, to represent is, ultimately, to intentionally select certain aspects of the target to stand for corresponding aspects of the source. In the case of scientific representation, models, broadly understood, are often the source of representation. Significant aspects of the reasoning that scientists engage when developing their research can be accommodated in terms of the framework just introduced. Scientists explore the consequences from the models they use, and even when the information is incomplete, which it typically is, they try to reason about the possible scenarios in terms of the resources offered by the models at hand.

But models are certainly not the only source of representation in scientific practice: micrographs, outputs of various instruments, templates, and diagrams, to mention just a few examples, are also commonly used. In the next section, I'll discuss this claim, by examining, in particular, one important kind of scientific representation: that provided by microscopy.

4. Microscopy, representation, and partial mappings

In which sense does microscopy offer a kind of representation? It might be argued that it doesn't. After all, the objection goes, once the sample is prepared and is suitably placed under the microscope it is not up to the researcher what kind of image will be obtained. Thus, we cannot say that there is an intentional component involved. And, as characterized above, without intentionality, there is no representation.

But this objection doesn't seem right. There certainly are intentional components involved in microscopy. In fact, a host of intentional moves are involved, for example, in the construction of the microscope to record certain interactions between the sample and the instrument, in the suitable preparation of the sample, and in the interpretation of the results.

By enhancing selected features of a sample, the microscope allows researchers to study components of that sample that otherwise would not be directly accessible to them. A significant part of the work is to distinguish artifacts of the instrument from genuine features of the sample. Various strategies are employed to this effect: from the use of different methods of sample preparation to the triangulation of the results obtained by a given microscope with other instruments (see, for example, Hacking, 1983).

Once the outcomes of a microscope are obtained, they can be taken to represent the features of the sample that were selected for study. This is achieved as the result of suitable partial mappings between the sample and the image that is generated by the instrument. The instrument was designed in order to be sensitive to certain features of the sample. If all goes to plan, when the microscope finally produces an image, the latter can be taken to represent the features of the sample that were selected for study. However, this doesn't settle the issue as to whether the representation is *accurate*. This is an additional issue. After all, even if the microscope is working properly, it's still possible that, unbeknown to the researchers, it fails to represent some aspects of the sample, or even misrepresents some of them in subtle ways.

To illustrate these points, I'll consider briefly two cases of representation in microscopy. I'll first examine the use of the transmission electron microscope (TEM) in the identification of ribosomes by George Palade. I'll then discuss the use of the atomic force microscope (AFM) in the identification of surface reconstruction in protein crystals. Both are typical cases of the use of the corresponding microscopes and both involve, as will emerge shortly, significant aspects of representation in microscopy.

In 1955, George Palade published a paper with the rather cryptic title: 'A Small Particulate Component of the Cytoplasm' (Palade, 1955). In this paper, Palade describes the new structure he identified at the cytoplasm by a careful and thorough use of the transmission electron microscope. At this point in the 1950s, the use of TEMs was already well established in biology, and the success of this instrument was, in part, supported by decades of systematic and effective use of optical microscopes in many areas of biology. Despite this success, Palade was very careful in determining what exactly were those 'small particulate components' he identified in a number of his micrographs. He first carefully excluded the possibility that they were just artifacts of the method of preparation, by changing the method and determining whether the 'small components' were still detected (they were). He then conducted additional research by using a centrifuge, and determined that there were indeed components with different densities in the sample. All of these results strongly supported the conclusion that the 'components' were indeed genuine (Palade, 1955).

Palade also conjectured in the paper that the 'components' in question were largely composed by ribonucleic acid, but he noted, additional chemical analysis was still needed to determine that. A few years later, the analysis did confirm Palade's conjecture, and the 'particulate components' were called 'ribosomes'.

When we consider the micrographs that Palade produced with his TEM, the presence of all those small particles at the cell's membranes is indeed remarkable. And it was largely based on the TEM images that Palade produced that the biological community very quickly accepted his results.

Do Palade's micrographs represent ribosomes? Clearly, Palade didn't simply make up the particular results he obtained with the TEM. He didn't create the microscopic images out of his own whim. Once the cell sample was suitably prepared,[5] and the instrument was applied to the sample, which micrograph Palade obtained was not up to him. This independence of the results from Palade's intentions supports the objectivity of the TEM. In this sense, the micrograph *doesn't* provide a representation of the ribosome, since there is no particular *intentional* relation involved. However, once the reliability of the TEM is established, *we* can use the micrograph as a representation of ribosomes, since we can take the small particles near the membranes of the cell in the micrograph to stand for the ribosomes. We might be mistaken, of course, in making such identification. And if this turns out to be the case, we can say that a misrepresentation was involved.

In terms of the framework for representation discussed above, a micrograph can be used as a representing device, where the targets are the ribosomes, and the sources are the corresponding 'small particulate components' on the micrograph. There is an indexical feature in the representing process, since in order to interpret the micrograph we need to locate ourselves with respect to it. We need to locate ourselves in relation to the 'small particulate components' on the micrograph, so that we are able to identify them, and determine what they stand for. We need to have some evidence to the effect that the 'small particulate components' that we are currently experiencing on the micrograph are not just some artifact of the preparation method, but do stand for some significant component of the cell structure. The evidence is unlikely to be conclusive, but if it's strong enough, it will end up supporting the adequacy of the representation. In this way, the micrograph can be interpreted as representing a novel cellular component – the ribosome. And it's in virtue of the features that support the adequacy of the representation that we are entitled to take the ribosome as a new component in the cell structure. The mechanism of representation, in this case, involves partial mappings from the micrograph to the sample that preserve the relevant structure – the geometrical configuration, location, and distribution of the 'small particulate components' on the micrograph.

Note that a scientific representation need not be true to be good. If it is quasi-true, and captures not every aspect of the sample, but only a limited selection of them, this is typically more than enough.

The TEM is an instrument for the study of the inner structure of the sample. In order to provide a representation of the inner structure of the sample, it ends up destroying the sample's surface. But if we are interested in studying the inner features of the sample, another type of microscope needs to be used, such as the atomic force microscope. The central idea of the AFM is that by scanning the surface of the sample, with a tip that is one atom sharp, information about the topography of the sample's surface can thereby be obtained (Binnig et al., 1986). The movements of the AFM tip are controlled by a cantilever, which gets the tip closer to, or moves it away from, the surface of the sample. The cantilever, in turn, is controlled by a laser beam, which makes the cantilever move up and down. As the tip scans the surface of the sample, it responds to the sample's topography, and delineates the 'contour' of the sample's surface. And if all goes well, this process will provide information about the geometric configuration of the atoms at the surface of the sample.

The AFM is, of course, a remarkable instrument. Together with the scanning tunneling microscope (STM) – which uses the quantum tunneling effect to establish a 'tunneling current' between the tip of the microscope and the sample, thus also obtaining information about the sample's surface (Binnig and Rohrer, 1983) – the AFM is a significant source of information about the surface of various materials in probe microscopy.

In 1999, using the AFM, a group of chemists at the University of Toledo made an important discovery (Li et al., 1999). The group was studying the geometrical configuration of atoms at the surface of protein crystals. They initially obtained information about the samples by a series of repeated measurements of their surfaces using the AFM. The multiple data obtained in this way were then statistically combined to produce an idealized imaged that offered the average of the various measurements that were made. This image is called the *experimental image*.[6]

Although the resolution of the experimental image is extremely impressive, it's not enough to settle several issues about the geometrical configuration of the atoms at the surface of the sample – in particular, the molecular-packing arrangement. To address this issue, the researchers create a *theoretical image*, in which they introduce (by hand, as it were) their hypotheses about what they take to be the particular atomic configuration at the sample's surface. The two images (the experimental and the theoretical) are then merged together, so that they can be compared. This produces a third, *hybrid image*, containing both the visual information from the experimental image and the hypothetical information from the theoretical image. Once the hybrid image was obtained, it became clear that there was a huge mismatch between the theoretical and the experimental images: the correlation between the two images was 62 percent. It wasn't strong enough.

The researchers then returned to the theoretical image, and made a novel hypothesis. The phenomenon of surface reconstruction – the difference between the packing arrangement of the surface of the crystal from the arrangement in the crystal interior – was well known in the case of inorganic crystals. The idea is clear: given that the surface molecules have incomplete bonds, and are in contact with the environment, these molecules tend to interact with the latter, thus changing the packing arrangement. However, it was unknown whether that phenomenon was also found in organic crystals – the subject of the investigation by the team. The researchers then changed the theoretical image by incorporating the assumption that surface reconstruction took place. The result was a new theoretical image, which was then compared with the original experimental image, thus obtaining a second hybrid image. The correlation now between the theoretical and the experimental images was 93 percent – a much more significant result.

Note how the representation process gets established in this AFM case. First, the target is the particular molecular-packing arrangement at the surface of protein crystals. The source is obtained in stages: from the experimental image, which provides more reliable, but partial, information about the molecular-packing arrangement, through the theoretical image, which provides more complete, but less reliable, information, to the hybrid image, which provides a combination of the two previous types of images. In each case, the researchers are dealing with *concrete* images, which are perfectly visible – even though they may have been generated by some statistical device. In each case, the researchers need to locate themselves with respect to the image, so that they can extract the relevant information about the molecular-packing arrangement. So, in each case, there is an indexical component in these representations. Moreover, in each case, the representation is established from a certain perspective, providing topographical information from the point of view of the tip of the microscope. And, once again, the mechanism of representation is provided by partial mappings from the various images to the sample that preserve the relevant structure – in particular, the geometrical configuration of the atoms at the sample's surface. And, in particular, a representation need not be true to be good. In fact, even in the best scenario, the correlation between the theoretical and the experimental image has never been perfect. A representation that preserves, via suitable partial mappings, the relevant aspects of the domain under study is often good enough.

Note also that the microscopic images are used as *inferential devices*. Given the poor correlation between the experimental image and the first theoretical image, the researchers drew information from a different domain (that of inorganic crystals) to produce a new theoretical image on the assumption that the phenomenon of surface reconstruction took place. This means that the researchers interpreted the protein crystals *as* going through a surface reconstruction, and made the necessary adjustments to, and drew suitable conclusions from, the theoretical image in light of the new interpretation.

Not surprisingly, microscopic images are often used as exemplars in the domain from which they emerge. They are taken to be typical of a certain group of phenomena, but they are also about a particular kind of sample. These images have both generality and specificity. They are general in that they are supposed to cover a variety of similar types of samples. They are specific in that they represent something particular about the instance of that type. Both features (generality and specificity) emerge from the way these images are constructed, in particular by using statistical techniques in the composition of the experimental data. The generality and specificity of the microscopic images allow researchers to use these images as inferential devices, so that they can generalize the information provided by a particular image to other samples (whether in the same domain or in related ones).

Finally, it's important to note the role played by *coding conventions*. To interpret a microscopic image, it's crucial to be able to identify the coding conventions that are in place – e.g. what do colors, shapes, and brightness in the image stand for? Unless coding conventions are specified, misunderstandings can easily emerge. It would be a serious mistake to think that if we could shrink ourselves to the size of an atom, and were wandering on the surface of the sample, we would observe exactly the same image as the one produced by the AFM.

5. Conclusion

In this chapter, I sketched a framework to accommodate some salient features of scientific representation. By making room for the fact that representation in scientific practice typically emerges in contexts where only partial information is available, the framework offers useful resources to make sense of actual cases of scientific representation. The framework also accommodates the diversity of representational mediums used in scientific practice, from models and templates to micrographs, images, and the output of various instruments. Moreover, since the framework offers a not so tight connection between the source and the target of representation – in terms of various partial morphisms – it can also accommodate the transfer of information even in contexts where the latter is incomplete.

There is, of course, much more to be said about these issues. But I hope that enough was said to motivate the framework and the prospects it has to illuminate interesting aspects of scientific representation.

Notes

1. For a fuller discussion of this issue, see Bueno (2005, 2008).
2. The partiality of partial relations and structures is due to the incompleteness of our knowledge about the domain under investigation. With additional information, a partial relation can become a full relation. Thus, the partialness examined here is not ontological, but epistemic.
3. For simplicity, I'll take the partial relations in the definitions that follow to be two-place relations. The definitions, of course, hold for any *n*-place relations.
4. A partial function is a function that is not defined for every object in its domain.
5. And here, as noted, considerable care goes into ensuring that the method of preparation does not generate the results – the latter are invariant under transformations of the preparation method.

6. It's interesting to note that, in their construction of the experimental image – which, as we saw, depicts the averages of multiple measurements – the AFM researchers continued a long-standing tradition in microscopy, stemming from Hooke's original presentation of his observations with the optical microscope. Instead of simply reproducing the results of individual observations, Hooke drew images that were meant to capture the average features of the many samples under study (see Wilson, 1995). A significant difference in the practice, from Hooke's work to the use of AFMs, is the introduction of computer software to produce the averages involved in the experimental image.

References

Berg-Hildebrand, A. and C. Suhm (eds) (2006) *Bas C. van Fraassen: the Fortunes of Empiricism* (Frankfurt: Ontos Verlag).

Binnig, C., C. Quate, and C. Gerber (1986) 'Atomic Force Microscope', *Physical Review Letters*, 56: 930–933.

Binnig, G. and H. Rohrer (1983) 'Scanning Tunneling Microscopy', *Surface Science*, 126: 236–244.

Bueno, O. (1997) 'Empirical Adequacy: a Partial Structures Approach', *Studies in History and Philosophy of Science*, 28: 585–610.

Bueno, O. (2000) 'Empiricism, Mathematical Change and Scientific Change', *Studies in History and Philosophy of Science*, 31: 269–296.

Bueno, O. (2005) 'Dirac and the Dispensability of Mathematics', *Studies in History and Philosophy of Modern Physics*, 36: 465–490.

Bueno, O. (2008) 'Scientific Representation and Nominalism: an Empiricist View', forthcoming in *Principia*.

Bueno, O. and E. de Souza (1996) 'The Concept of Quasi-Truth', *Logique et Analyse*, 153–154: 183–199.

Bueno, O., S. French, and J. Ladyman (2002) 'On Representing the Relationship between the Mathematical and the Empirical', *Philosophy of Science*, 69: 497–518.

da Costa, N. C. A. and S. French (2003) *Science and Partial Truth* (New York: Oxford University Press).

French, S. and J. Ladyman (1997) 'Superconductivity and Structures: Revisiting the London Account', *Studies in History and Philosophy of Modern Physics*, 28: 363–393.

French, S. and J. Ladyman (1999) 'Reinflating the Semantic Approach', *International Studies in the Philosophy of Science*, 13: 103–121.

Hacking, I. (1983) *Representing and Intervening* (Cambridge: Cambridge University Press).

Levine, G. (ed.) (1993) *Realism and Representation* (Madison: University of Wisconsin Press).

Li, H., M. A. Perozzo, J. H. Konnert, A. Nadarajah, and M. L. Pusey (1999) 'Determining the Molecular-Packing Arrangements on Protein Crystal Faces by Atomic Force Microscopy', *Acta Crystallographica*, D55: 1023–1035.

Mikenberg, I., N. C. A. da Costa, and R. Chuaqui (1986) 'Pragmatic Truth and Approximation to Truth', *Journal of Symbolic Logic*, 51: 201–221.

Palade, G. (1955) 'A Small Particulate Component of the Cytoplasm', *Journal of Biophysical and Biochemical Cytology*, 1: 59–79.

Sigman, J. and B. van Fraassen (1993) 'Interpretation in Science and in the Arts', in G. Levine (ed.), *Realism and Representation* (Madison: University of Wisconsin Press), 73–99.
van Fraassen, B. (1980) *The Scientific Image* (Oxford: Clarendon Press).
van Fraassen, B. (2006) 'Weyl's Paradox: the Distance Between Structure and Perspective', in A. Berg-Hildebrand and C. Suhm (eds), *Bas C. van Fraassen: the Fortunes of Empiricism* (Frankfurt: Ontos Verlag), 13–34.
Wilson, C. (1995) *The Invisible World* (Princeton: Princeton University Press).

6
The Identical Rivals Response to Underdetermination

Gregory Frost-Arnold and P. D. Magnus

The underdetermination of theory by data obtains when, inescapably, evidence is insufficient to allow scientists to decide responsibly between rival theories. One response to would-be underdetermination is to deny that the rival theories are distinct theories at all, insisting instead that they are just different formulations of the same underlying theory; we call this the *identical rivals response*. An argument adapted from John Norton suggests that the response is presumptively always appropriate, while another from Larry Laudan and Jarrett Leplin suggests that the response is never appropriate. Arguments from Einstein for the special and general theories of relativity may fruitfully be seen as instances of the identical rivals response; since Einstein's arguments are generally accepted, the response is at least sometimes appropriate. But when is it appropriate? We attempt to steer a middle course between Norton's view and that of Laudan and Leplin: the identical rivals response is appropriate when there is good reason for adopting a parsimonious ontology. Although in simple cases the identical rivals response need not involve any ontological difference between the theories, in actual scientific cases it typically requires treating apparent posits of the various theories as mere verbal ornaments or computational conveniences. Since these would-be posits are not now detectable, there is no perfectly reliable way to decide whether we should eliminate them or not. As such, there is no rule for deciding whether the identical rivals response is appropriate or not. Nevertheless, there are considerations that suggest for and against the response; we conclude by suggesting two of them.

1. Responding to underdetermination

Consider a *prima facie* case of underdetermination: there seem to be two theories T_1 and T_2 such that standards of responsible theory choice preclude deciding between them. If we accept that the case is as it seems, then we must decide what to do about T_1 and T_2. A modest *agnostic* response is to suspend judgement. If our actions depend on the difference between them, then we muddle along in our uncertainty. A more bumptious response – call this *fideism* – is to believe one of the two theories anyway, as an article of faith.[1]

Of course, since this is only an *apparent* case of underdetermination, we might instead try to show that the choice is not really underdetermined at all. There are again two possibilities. We might deny that responsible choice between the two is impossible. If we tinker with the standards of theory choice, then there might be grounds to prefer one or the other. This is a standard gambit of realists, who add simplicity, novel prediction, or explanatory power to the standards that guide theory choice. To highlight the contrast with agnosticism, call this the *gnostic* response.

Alternately, we might deny that there really are two rival theories under consideration. This might seem as bumptious a response as fideism, but it need not be. Imagine, for example, that T_1 is a theory in French and that T_2 is its German translation; that T_1 is a theory expressed in the metric system and that the only difference in T_2 is that values are converted to imperial units; or that T_1 is formulated using Cartesian coordinates and that T_2 expresses equivalent claims using polar coordinates. Surely, standards of responsible theory choice will be insufficient to prefer one over the other with any of these three pairs. Within each pair, both of the options are the selfsame theory! One merits belief just if the other does. There really is no underdetermination at all. We might reply to any apparent case of underdetermination by insisting that the alleged rivals are really just different formulations of the same theory. Call this the *identical rivals* response.

These four responses form a matrix:[2]

	We do not decide between T_1 and T_2	*We do decide between T_1 and T_2*
The case is judged to be underdetermined	agnostic response	fideist response
The case is judged not to be underdetermined	identical rivals response	gnostic response

For positivists, the identical rivals response is always the right one: Any two theories which cannot be distinguished on observational grounds have the same meaning and so are the same theory. Positivism runs into familiar difficulties – and there are probably no positivists among our readers, anyway – so we set that answer aside. What we want to know is, under what circumstances is the identical rivals response appropriate?

The obvious answer is: when the two theories *really are* identical. The difficult and interesting question is, rather, what counts as good evidence for the identity of two apparently inconsistent but observationally equivalent theories? Put otherwise, what sorts of considerations would lead one to view two such theories as identical? We propose to come at this difficult question in a roundabout way. In section 2, we consider famous Einsteinian arguments that make use of the identical rivals response. These examples show that the response is appropriate in at least some cases. Moreover (as we discuss in sections 3–4) they indicate that the identical rivals response goes naturally with a parsimonious ontology. We argue further: the identical rivals response is really only tenable in cases where a parsimonious ontology is tenable. In section 5, we use this connection to provide some criteria for when the identical rivals response is appropriate.

2. Two examples from Einstein

Einstein employs what is recognizably the identical rivals response in formulating both the special and general theories of relativity. The same inferential pattern appears in his 1905 special relativity paper and reappears, applied to different subject matter, in his justification of the principle of equivalence, one of the essential steps on the intellectual road to general relativity.

Einstein (1905/1952) claims that Maxwell's electrodynamics suffers from certain faults. He presents these faults via the now-famous example of a magnet and a conductor moving relative to one another. If the magnet is regarded as moving and the conductor as being at rest, Maxwell's theory claims that an electric field with a certain energy will come into existence around the magnet, which produces an observable current in parts of the conductor. However, if the magnet is regarded as being at rest and the conductor as moving, then there is no electric field in the neighborhood of the magnet. Instead, an electromotive force is created in the conductor. This electromotive force gives rise to the very same current in the conductor as in the first case. If we move with the magnet, we say that an electrical field is produced around the magnet; if we move with the conductor, we say an electromotive force is produced in the conductor. In both cases, however, the current running though the conductor is the same; if we connected an ammeter to the conductor, we would measure the same value for the current in both cases. Einstein draws the following moral from these two cases: 'Maxwell's electrodynamics ... when applied to moving bodies, leads to asymmetries which do not appear in the phenomena' (1905/1952: 37).

Let T_1 be the description according to which the magnet is in motion and the conductor is at rest; let T_2 be the description according to which the conductor is in motion and the magnet is at rest. T_1 and T_2 make the same predictions for all observable phenomena. We have a *prima facie* case of underdetermination.

In a similar case, Galileo counseled an agnostic response. In his famous discussion of a moving boat, Galileo concludes that the people below decks cannot tell whether they are in motion or not – it is 'as if' the people are at rest (1632/1967: 116). Given two bodies in relative motion, Galileo would have said, there is a fact of the matter about whether one of them is at rest or not – even though no local appearances could allow us to determine which of the two is moving. In this case there is a genuine agnostic response to putative underdetermination; it is not merely a conceptual possibility marked out in the matrix above.

Yet Einstein does not give Galileo's agnostic response. Rather, he concludes that T_1 and T_2 are just different descriptions of the same situation. Although an orthodox physicist in 1905 would think that either the magnet or the conductor was in true motion, the supposed difference between two descriptions is illusory. In short, Einstein gives the identical rivals response.

Einstein deploys the identical rivals response when arguing for the general theory of relativity as well. His reasoning is especially clear in the first general relativity seminar he taught, in the summer semester of 1919 in Berlin. Hans Reichenbach attended this seminar, and his notebook contains formulations of Einstein's that bring out the parallels to the 1905 paper. Early in the term, Einstein states that both classical mechanics and the special theory of relativity suffer from a certain 'physical deficiency'. Both rely on the natural law that gravitational mass equals inertial mass, which has been confirmed 'with very great, astounding exactness' (1919: I.3). Nevertheless, classical mechanics must take the equality as a brute fact and consequently 'there is no explanation for ... [this] most important law of nature'. As Einstein sees it, though, 'we want the numerical equality reduced to an essential equality' (1919: I.4). Einstein accounts for this equality by the so-called *principle of equivalence*.[3]

Einstein considers two frames of reference. The first frame is inertial, i.e. it has zero acceleration in all three spatial dimensions, but has a homogeneous gravitational field of strength g directed along the y-axis in the negative direction.[4] The second frame is non-inertial, for it accelerates in the y-direction at the rate γ, but has zero acceleration along the other axes. The equations of motion for an observer at rest with respect to the second frame 'are the same equations that describe motion in the [homogeneous] gravitational field. We can thus say: [the second frame of reference] is at rest, but a gravitational field is present. We need only set $g = -\gamma$. Through this conception, the essential difference between inertial and heavy mass is taken away' (1919: I.6).

How should one interpret the physical meaning of the equality $g = -\gamma$? In Reichenbach's notebook, there is a diagram of the two frames. Parallel to the y-axis in each, there is a drawing of a spring. Beneath the drawing, Reichenbach has written: 'From the point of view of [the second frame of reference]: the tension of the spring arises through *gravitational* mass. From the point of view of [the first frame of reference], one judges: The tension of the spring arises through the inertial resistance of the body, through *inertial* mass. The same effect, in one case from inertial mass, in the other case from gravitational mass' (1919: I.6) but it is interpreted differently in different frames of reference.

The parallel with the 1905 inference is close to the surface. If the magnet is regarded as resting, then an electromotive force arises in the conductor; while if the conductor is regarded as resting, then an electrical field arises around the magnet. In either case, we observe the same value of the electrical current. Similarly, in the general relativistic case, if the system is considered to be in motion (constant acceleration) in a gravitation-free region, then the spring experiences inertial resistance; while if the system is considered to be at rest in a uniform gravitational field, then the spring experiences weight due to gravity. In both cases, a spring-based scale would measure the same force. Given the meter reading for the spring, we can explain this reading by positing an inertial force acting on it (the scale is accelerating in a gravitation-free region), or a gravitational force acting on it (the scale is at rest in a uniform gravitational field). One could accept this underdetermination at face value and respond with agnosticism, holding that one or the other description is correct even though we limited beings cannot say which. Einstein does not respond in this way. Rather, he concludes that the two forces are in fact 'essentially' the same, so that there is no real difference between an accelerated spring in a space free of gravitational forces and a spring at rest in a homogeneous gravitational field.

The difference between Einstein's two cases stems from the difference between the special theory, which maintains the notion of a privileged set of inertial frameworks, and the general theory, which attempts to do without such structures. The *constant velocity* of the 1905 paper is replaced by *constant acceleration* later. But both of these arguments pose would-be underdetermination scenarios between two descriptions, and defuse the underdetermination by way of the identical rivals response.

3. Ontology and the examples

The previous section illustrated two instances of the identical rivals response. They are sufficient to show that the identical rivals response is at least sometimes, if not always, appropriate – at least for philosophers of science who consider Einstein's arguments to be good.

Nevertheless, both go beyond the identical rivals response as we originally posed it. The trivial case was when we imagined the same theory in French and in German (or in metric and imperial units of measure, or in Cartesian and polar coordinates) – suppose, for the sake of concreteness, that we have contemporary biochemistry in two languages. The theories talk about a great many unobservable things, like amino acids, enzymes, and so on, as well as their many properties and relations. When we make the identical rivals response, we accept each of the theories. We accept that there really are amino acids and whatnot. For present purposes, it does not matter if this is the practical acceptance of a constructive empiricist or the belief of a realist. What matters is that the two theories are understood as positing a rich ontology of unobservable stuff. We accept the ontologies of the two theories at face value, and – as part of the identical rivals response – we insist that the two ontologies are the same.

The two non-trivial examples are importantly different than this. Consider Einstein's first argument. If we treat the two representations as making claims about absolute motion and absolute rest, then certainly they disagree. Moreover – since there is no absolute motion – neither description is true.[5] It is crucial that Einstein does not treat them in this way. He is only able to treat them as different descriptions of the same situation because he prunes their ontological commitments. There is no absolute motion, only relative motion. Einstein's second argument is similar. In employing the identical rivals response, he prunes any difference between constant acceleration and gravitation from his ontology. The distinctions drawn in the earlier theories are merely verbal distinctions without a difference.

So each extended example of the identical rivals response involves some ontological parsimony; that is, it involves eliminating some of the things that naïvely seem to be posited by the two theory formulations. Yet, as the trival examples are sufficient to show, the identical rivals response is not of necessity eliminativist. So is this just an accidental feature of the examples from Einstein that we've discussed? We argue that it is not. The identical rivals response, except in trivial cases (such as French–German and metric–imperial), requires ontological parsimony.

In the next section, we attempt to deliver on this claim.

4. General arguments

Our approach in this section begins with a sort of antinomy: an argument by John Norton purports to show that (to put it in our terminology) the identical rivals response is presumptively appropriate whenever we seem to be faced with empirically equivalent theories. At the other extreme, an argument by Larry Laudan and Jarrett Leplin purports to show that there could never be empirically equivalent theories. A consequence of the latter argument is that whenever we seem to be faced with empirically equivalent theories we must be mistaken; as such, the identical rivals response would never be appropriate. So these arguments seem to reach incompatible conclusions. This antinomy is resolved by explicitly acknowledging the role of ontological parsimony in applications of the identical rivals response. Before arguing for this, we should explain the two arguments.

THESIS: *Anytime we are faced with apparent underdetermination, the identical rivals response is probably appropriate.* To give the argument for the thesis briefly:

> Suppose we have two theories which are demonstrated to have the same observational consequences.
>
> Since the observational consequences can be reasoned about in this way, there must be some tractable description of them. If the observational consequences of a theory can be described compactly without recourse to the non-observable posits of the theory, then the posits are otiose. So we may presume this not to be possible. There are descriptions of the observational consequences of each theory that make essential use of the central theoretical terms of that theory.
>
> There are three possibilities: either the theoretical structures of the two theories are utterly distinct, one theory has surplus structure, or they are interconvertible without loss.

> (1) The theoretical structure of each theory is what systematizes its observational consequences. As such, the demonstration that the theories have the same observational consequences must have exploited some similarity between the theoretical structures of the two theories. So we may conclude that the two theories do not have utterly different theoretical structures.
> (2) If one theory has surplus structure, then – since the rival theory produces the same observational consequences without the surplus – the structure must be inessential for generating observational consequences. These are otiose. They are (as Norton says) 'strong candidates for being superfluous, unphysical structures' (2008: 35).
> (3) If the theoretical structures are interconvertible without loss, then we should think that the two theories are really just saying the same thing.
>
> Since the third possibility is the most likely, we have a strong reason to think that the two are merely different formulations of the same theory.

Norton puts the conclusion of the argument this way: '[P]airs of theories that can be demonstrated to be observationally equivalent are very strong candidates for being variant formulations of the same theory' (2008: 35). As we would put the point: the identical rivals response is probably appropriate for any case in which the theories can be proven to have the same observational consequences.

Norton's formulation of the argument is somewhat weaker than what we have in mind. It 'is specifically restricted to those [theories] whose observational equivalence can be demonstrated in the sort of compact argumentation that can appear in a paper in the philosophy of science literature' (2008: 33). And there may be a gap between a case being a 'very strong candidate' for the identical rivals response (as Norton says) and the response's being probably appropriate. Regardless, the same argument form can be redeployed to yield the conclusion that the identical rivals response should be the presumptive reaction to empirically equivalent theories, and we will refer to this as 'Norton's argument' below.

Norton's argument raises several issues, but we wish to concentrate on his answer to the second possibility: one of the theories has surplus structure that is not essential for producing the theory's observational consequences. Toy examples of this kind are easy to produce. Let one theory be standard particle physics, and let another theory be standard particle physics plus the posit that there is an undetectable dragon in my garage. Of course the observational consequences of these two theories are the same, and one cannot argue on the basis of observation that there is no dragon. Yet it is tempting to treat the dragon not as an actual posit of the second theory, treating it instead as 'superfluous' and 'unphysical'.

However, refusing to treat dragons as a serious posit of the second theory is not entirely innocent. Imagine that a century from now there is a technique for detecting previously undetectable dragons. Future scientists might come to the place where my garage once stood, turn on their dragonometers, and decide between these two theories. The invisible dragon would not be superfluous or unphysical after all. Returning to the more realistic cases of relativistic physics, it could turn out that future scientific generations will find good reasons to posit absolute velocities or to distinguish inertial effects from gravitational ones. For example, Bohm's interpretation of quantum mechanics requires absolute velocity. If Bohmian mechanics is eventually accepted, then what Einstein considered surplus structure in classical electrodynamics will once again play an important theoretical role. Similarly, if future technological developments allow for more precise versions of Eötvös's experiments to be carried out, then it is conceivable that inertial and gravitational forces will need to be pulled apart again, overturning Einstein's identification of them in the principle of equivalence; see, for example, Einstein (1922/2002: 316). The generality of the considerations raised in this paragraph suggest:

ANTITHESIS: *When faced with apparent underdetermination, the identical rivals response is never appropriate.* Laudan and Leplin (1991) argue that, in general, there cannot be logically distinct[6] but empirically equivalent theories. Their reasoning is similar to that of the previous paragraph. To summarize briefly:

> The boundaries of the observable are historically variable; similarly, the auxiliary hypotheses that scientists employ in making predictions are historically variable. So take two theories that only disagree about matters that are *now* unobservable. We cannot rule out the possibility that at some time in the future we will develop a way to observe the relevant differences; similarly, scientists might learn previously unknown auxiliary laws which connect the previously unobservable differences with observable consequences. As such, we cannot say of two such theories that they are empirically equivalent.

In the face of any putative underdetermination, we should deny that the rivals are empirically equivalent. This precludes employing the identical rivals response, because if they are the same theory then there cannot be any empirical difference between them.

If the supposition required at the outset of Norton's argument were timeless empirical equivalence with no logical possibility of observational discrimination, then Laudan and Leplin's argument would suffice to show the untenability of Norton's thesis: even though two theories may appear observationally equivalent given today's state of the art, tomorrow's unforeseeable new discoveries may overturn this apparent equivalence. However, Norton's argument need not suppose anything so strong. Distinct theories can have the same observational consequences, *given* background assumptions about observability.[7] For the purpose of Norton's argument, we can treat physics-*cum*-dragon and physics-*sans*-dragon as having the same observational consequences, because we presume as tacit background knowledge that nothing remotely like a dragonometer exists. Dragonometers, as a very remote possibility, can reasonably be set aside as science fiction. On a charitable interpretation, Norton's argument would take the actual physical posits of the two theories as *probably* identical. If the 'dragons' mentioned in one formulation are unphysical posits, then there is no possibility of their detection.

Refusing Norton's suggestion and insisting that the dragons might one day be observable would require a literal reading of the dragon ontology, but parallel scruples would undo the Einsteinian examples (from section 2). Consider Einstein's argument that opens the 1905 special relativity paper. Just as we can imagine dragonometers, we can imagine übergyroscopes that could tell the difference between real motion and absolute rest, and thereby decide which of the two empirically equivalent descriptions is the true one.[8] Or as mentioned above, Bohmian mechanics, which requires bodies to have absolute velocities, could eventually be accepted. If there were an ontological difference between moving the magnet and moving the conductor, then there might in principle be some way to distinguish between the two. If one accepts Laudan and Leplin's argument, then the situation (in Einstein's 1905 argument) in which the magnet is at rest would *not* be empirically equivalent to the situation in which the magnet is in motion – because imaginable devices and theories could tease them apart. However, the two competing claims are still underdetermined, in the sense that we cannot responsibly decide between them, since we have no übergyroscope.[9] Similar considerations apply to differentiating inertial mass and gravitational mass. If one follows Laudan and Leplin, then our inability to pull these apart would be merely a brute fact, an observed 'numerical equality', as opposed to an 'essential equality'. They offer a version of what we earlier called the 'gnostic' response to apparent underdetermination – provided such fantastic devices, we could decide between the theories – but the two examples from Einstein show that the gnostic response is not always appropriate.

Bohmian mechanics and an übergyroscope both presuppose that there is an absolute difference between motion and rest; they take the ontologies of the two descriptions to be thoroughly physical, as opposed to 'unempirical' or 'superfluous'. The identical rivals response proposes instead that there is no such thing as absolute rest, so there is obviously no possible device for detecting it. Switching to Einstein's argument for the principle of equivalence, the identical rivals response is only appropriate because the 'numerical equality' of gravitational and inertial mass is replaced with identity – what Einstein might have called 'essential equality'. The plausibility of Einstein's eliminativist ontology is the very reason that his identical rival responses are not undone by Laudan and Leplin's considerations. Because the alternatives describe the same situation, they are logically (or at least metaphysically) equivalent and so necessarily empirically equivalent. If the eliminativist ontology is correct, there is no possible future development that could make them yield different predictions.

Returning to Norton's argument, his answer to the second possibility is crucial. Ontological parsimony – refusing to take the excess structure as physical – rules out the possibility that the excess part will generate detectable differences as science advances. It is important to note that this eliminativist move is typically a crucial part of deploying the identical rivals response. We treat undetectable dragons as superfluous because the rest of science directs us to dismiss squamous phantoms. Yet we are fallibilists, and we recognize that we might be wrong; so we cannot banish dragons completely – dragons and absolute velocities remain (in a weak sense) epistemically possible. If scientists in a century develop dragonometers or übergyroscopes, our parsimonious ontology and identical rivals response will prove wrong in retrospect. Norton's argument only yields the conclusion that surplus structures are 'candidates' for occamizing; any two theories with the same observational consequences are candidates for the identical rivals response. Whether the identical rivals response should be elevated from mere candidacy to full adoption depends crucially on whether there is good evidence for ontological parsimony.

Here is another way of making the same basic point. Parties on both sides of the antinomy can accept the conditional 'If two theories are truly empirically equivalent, then they are identical simpliciter'. The thesis (Norton's conclusion) follows from accepting the antecedent and, by modus ponens, deriving identity. The antithesis (Laudan and Leplin's conclusion) follows from rejecting the consequent and, by modus tollens, deriving their empirical inequivalence. We have attempted to show in this section that both modus ponens and modus tollens, as uniform policies, would be rash when facing apparent cases of underdetermination. The success of Einstein's gambits shows that Laudan and Leplin's modus tollens view is too strong, while Laudan and Leplin's plausible point concerning the variability of auxiliary hypotheses, and resultant variability of what is observable, shows that modus ponens would be too strong. An important reason why uniform modus ponens and uniform modus tollens are both too extreme is that the identical rivals response typically involves (in realistic cases) ontological eliminativism, and there is simply no rule to always eliminate or never eliminate – or even a default rule to eliminate or not. As philosophers should be well aware from debates over several specific eliminativist proposals, the evidential situation pro or contra elimination is usually rather subtle and complex.

5. Two conditions, but no rule

The discussion so far shows that the identical rivals response is sometimes appropriate but cannot necessarily be applied to all cases of apparent underdetermination. Applying it is not simply a matter of inspecting the meanings of the would-be rival theories. It is usually a matter of *deciding* how to understand them: treat the differences as substantive physical disagreement, or treat the differences as merely verbal. We strongly suspect that there is no determinate algorithm for this decision. We can never with absolute certainty rule out the possibility that future auxiliary theories might change the observable consequences of the theories, making today's superfluous content into tomorrow's well-confirmed posit. The point here is not just that such an outcome is logically possible. The Laudan–Leplin worries can always be formulated so long as the two rival theories are treated as genuinely distinct. As such, whether the would-be rivals are merely variant formulations of the same theory cannot be known with certainty. Despite the reasonable doubts raised by the Laudan–Leplin arguments, the two examples from Einstein are sufficient to show that the identical rivals response is sometimes appropriate.

Even though there is no algorithm, we want to suggest two kinds of considerations that are relevant to this decision.

5.1 The future discriminability condition

The identical rivals response is ultimately untenable if future developments will allow scientists to observationally distinguish between the rival theories. If future technology can detect absolute motion, then it is a mistake even now to treat the disagreements about absolute motion as merely verbal differences. Conversely, the identical rivals response is tenable if such future developments will not occur. Call this the *future discriminability condition*, because it is not directly available to deliberation in the present moment. It is always a matter of whether future possibilities will be realized or not.

However, there can be directly available evidence for or against the future development of observable differences. Before the first neutrino detectors, it would have been rash to trim them presumptively from our ontology; scientists reasonably expected that it would eventually be possible to detect neutrinos, as technology advanced. In cases like these, it is inappropriate to apply the identical rivals response.[10] Such cases contrast with e.g. absolute velocity, whose inaccessibility to our senses has nothing to do with its being too small for our visual system to detect. No matter how powerful our microscopes become, detecting absolute velocity will not become any easier. It would require an instrument that works in utterly novel ways, instead of simply extending our already-existing discriminatory powers. So the identical rivals response is more defensible here than in the case of neutrinos.

5.2 The heuristic utility condition

Independently, surplus structure may serve a useful heuristic role. Fundamental particles are often posited first only for systematic and theoretical reasons. They may be treated as genuine physical posits of the theory for some time before techniques are developed for actually observing them. Call this the *heuristic utility condition*: the identical rivals response is inappropriate when the peculiar posits of a theory are heuristically useful and guide scientists in developing the theory. Conversely, the response is appropriate when the posits serve no useful heuristic role.

A posit might be heuristically useful in this way even if it never makes an observable difference. Even if a particle theory is revised before techniques to observe the particle are developed, taking the posit seriously could be crucial for developing the theory; descendants of the theory may be ultimately confirmed.

In the remainder of this section, we consider two examples in which the internal condition plays a prominent role: quantum mechanics in 1926 and the indeterminacy of translation.

In the mid-1920s, Erwin Schrödinger's wave mechanics and Werner Heisenberg's matrix mechanics could have seemed like rival theories. In wave mechanics, the state of the system changed with time and the operators were time independent. In matrix mechanics, conversely, the state was time independent and the operators were time dependent. Nevertheless, any predictions of either would require both the state and an operator, and the predictions of each were (in general) time dependent. So it was natural to think that the disagreement is one about where to write the variable t, rather than a substantive disagreement about the world. Indeed, following papers by Schrödinger and Carl Eckart in 1926, physicists came to treat wave and matrix formulations of quantum mechanics as different formalisms for the same theory. As we would put it, they applied the identical rivals response.

Yet, as Muller (1997) has shown, Schrödinger's arguments were insufficient to show that the two were equivalent. Wave and matrix mechanics did not even have the same observational consequences at that time! So the two could not have been formulations of the same theory. Nevertheless, treating them as if they were was fruitful for physicists. What we are calling the internal condition – heuristic fruitfulness – vindicates the choice. Less than a decade later, von Neumann was able to represent the two approaches in terms of Hilbert spaces. Even though the theories were different in their details in 1926, *successors* of the two theories were just different expressions of a more general formalism.

In this case, as with the other cases we have considered, the identical rivals response involved ontological streamlining. Schrödinger's version of quantum mechanics described unobserved particles as fluctuating bits of jiggly matter; in Muller's phrase, 'tiny jelly-like lumps of vibrating charged matter' (1997: 229). Treating wave mechanics as equivalent to the anti-metaphyscial matrix mechanics meant treating these waves as unphysical.

Consider a different example: Quine's thesis of the indeterminacy of translation is a reaction to a putative underdetermination scenario.[11] Quine asks us to imagine two linguists working independently to translate the totally foreign Jungle language. Each creates a manual for translating between Jungle and English. The two might create manuals which prove equally serviceable in interactions between Jungle speakers and English speakers but which nevertheless disagree about which Jungle words correspond to which English words. '[T]he thesis of indeterminacy of translation' is that the two 'manuals might be indistinguishable in terms of any native behavior that they gave reason to expect, and yet each manual might prescribe some translations that the other translator would reject' (1987: 8). Considered as theories about the meanings of words in Jungle, the two translation manuals are empirically equivalent but incompatible theories.

This putative case of underdetermination might be met in any of the four general ways that we discussed in the introduction. If we took the agnostic route, we would infer that the existence of the two translation manuals shows that we English speakers cannot know which translation of a given utterance in Jungle provides the true meaning of that utterance.

Quine instead gives the identical rivals response: the existence of alternative translation manuals shows not that we should suspend judgement on which is the 'correct' or 'true' translation, but rather that neither is the one true or correct translation. As Quine says:

> The problem is not one of hidden facts ... The question whether ... the foreigner *really* believes A [the translation according to the first manual] or believes rather B [the second manual's translation], is a question whose very significance I would put into doubt. (Quine, 1970: 180–181)

Although the two manuals seem to disagree about the meaning of Jungle words and sentences, this appearance results only from thinking that the two manuals are theories about traditional (i.e. extra-behavioral) meanings of words or sentences. Quine defuses the would-be underdetermination by suggesting that there is no such thing as the traditional meaning of a sentence in isolation, just as Einstein concluded in 1905 that there is no such thing as absolute velocity. Different manuals are analogous to different inertial frames of reference.

If one is going to resist the Quinean deflation of meaning, it will not be by invoking the future discrimination condition. No one imagines that future linguists will be able to construct meaning-o-meters that determine which translation manual gives the true meanings of the words. Rather, it must be that there is some theoretical advantage to positing meanings. Katz (1967, 1997), for example, argues that such a theoretical advantage exists – that there is a systematic reason internal to linguistics or semantics for positing meanings.

This is a further illustration of the kind of consideration we have in mind with the heuristic utility condition: if there is no heuristic or theoretical advantage of one translation manual over the other, then there seems no ground left for resisting the identical rivals response.

6. Conclusion

We have argued for two general conclusions regarding the identical rivals response in actual scientific cases: first, it goes along with deflating the ontology of one or both theory formulations. The would-be rivals are made to agree because the points on which they could disagree are taken to be merely verbal differences. Second, the identical rivals response is never necessitated by the situation itself. It is, rather, a decision under uncertainty. It is a strategic choice about how to respond to apparent underdetermination.

In the previous section, we suggested two conditions which ought to influence this strategic choice. The future discriminability condition suggests that one should not apply the identical rivals response if there is evidence that future observations will allow you to detect effects of the posits about which the theories seem to disagree. The heuristic utility condition states: don't apply the identical rivals response if the extra features of one formulation provide extra heuristic or theoretical resources for developing the theory. (Although this formulates the conditions as negative ones, we do not mean to suggest that they merely operate as roadblocks to the presumptive application of the identical rivals response. The conditions could be worded positively, in terms of when the response ought to be applied.)

Note that applying the future discriminability condition depends on our background knowledge. Note also that the heuristic utility condition depends on our present theoretical context; whether something is fruitful or heuristically useful depends on what else we have to work with. As such, both of these conditions depend on time and context.

Even acknowledging that Einstein was right to apply the identical rivals response and treat absolute velocity as an unphysical posit, it does not follow that the response would have been appropriate earlier. For Newton, absolute velocity played an important role. He could not see how mechanics could proceed without it,[12] so what we have called the heuristic utility condition might have justified his rejecting Einstein's arguments (if, anachronistically, he had considered them). In the mid-nineteenth century, scientists might reasonably have hoped that the ether would provide a medium in which to measure the Earth's absolute velocity; the future discriminability condition would have justified their rejecting Einstein's arguments (again, if they had anachronistically considered them). It is hard to say when the time was right for Einstein's arguments, but certainly it had arrived by the time he proposed them.

This only seems shocking if we imagine that applying the identical rivals response is like intuiting a timeless essence. Yet it is not as if Einstein had peered into the Form of each state description and recognized them to be the same. Einstein made a scientific argument, situated in the scientific and evidential context of his time. Applying the identical rivals response was appropriate only given the background knowledge and theories of the time. And so it is in any case.

Notes

1. Regarding fideism, see Magnus (2005a) and also van Fraassen (2002).
2. Sklar (1974: ch. 2) and Gardner (1976) offer a similar assessment of the options.
3. Einstein expresses this idea in print. He writes that 'previous mechanics had indeed registered this important law, but had not interpreted it'. An acceptable 'interpretation' of this identity requires recognizing that one and 'the same quality of a body expresses itself, according to circumstances, as "inertia" or as "weight"' (Einstein, 1917/1997: 40). And elsewhere: 'It is ... clear that science is fully justified in assigning such a numerical equality only after this numerical equality is reduced to an equality of the real nature of the two concepts' (1922/2002: 56–57).
4. A gravitational field (in a space) assigns to every point in the space a gravitational vector; this vector is interpreted physically as the instantaneous acceleration experienced by a test particle at that point in the space. A gravitational field is homogeneous if and only if the same vector is assigned to every point in the space; obviously, the gravitational fields in the vicinity of the sun, the earth, or any other body are not homogeneous.
5. One might say that they are false, or that they lack truth values because of presupposition failure or some other semantic problems. But false and truth-valueless theories are both unacceptable, so this difference is unimportant here.
6. Depending on how theories are individuated, this should perhaps be 'metaphysically distinct'. For the terms 'inertial mass' and 'gravitational mass' are not *logically* identical, even for Einstein – however, they are necessarily or 'essentially' identical for Einstein: a sentence of the form $\forall x(Px \equiv Qx)$ is not a logical truth, but it can be true in all possible worlds for certain values of P and Q.
7. Of course, as Laudan and Leplin note, such assumptions are linked to auxiliary hypotheses that are themselves historically variable.
8. Indeed, responding to van Fraassen's discussion of absolute space, Laudan and Leplin imagine discoveries that are tantamount to an übergyroscope (1991: 458).
9. Although 'underdetermination' and 'empirical equivalence' are sometimes used interchangeably, this case would seem underdetermined even though

the theories would not be empirically equivalent. For another such case, see Magnus (2005b).

10. This position does not beg the question of realism against the constructive empiricist. By ruling out the identical rivals response in such cases, we are only insisting that the theories should be understood to actually posit neutrinos. The realist and the constructive empiricist agree on that much, and we leave them to argue over whether we should believe in the posits or merely accept as empirically adequate the theory that posits them.
11. Quine himself recognizes differences between the 'standard' underdetermination of theory by evidence in the sciences and his radical translation scenario. 'The indeterminacy of translation is not just an instance of the empirically underdetermined character of physics' (1970: 180). We are not concerned here with the nitty-gritty of Quine exegesis.
12. See especially Newton's unpublished manuscript *De Gravitatione et Aequipondio Fluidorum* (1962: II.1; in particular 129–131).

References

Einstein, A. (1905/1952) 'On the Electrodynamics of Moving Bodies', in *The Principle of Relativity*, trans. W. Perrett and G. B. Jeffery (New York: Dover).

Einstein, A. (1911/1952) 'On the Influence of Gravitation on the Propagation of Light', in *The Principle of Relativity*, trans. W. Perrett and G. B. Jeffery (New York: Dover).

Einstein, A. (1917/1997) *On the Special and General Theory of Relativity (A Popular Account)*, trans. R. Lawson, in *The Collected Papers of Albert Einstein*, Volume VI (Princeton: Princeton University Press).

Einstein, A. (1922/2002) *The Meaning of Relativity*, trans. E. Adams, in *The Collected Papers of Albert Einstein*, Volume VII (Princeton: Princeton University Press).

Galilei, G. (1632/1967) *Dialogue Concerning the Two Chief World Systems*, trans. S. Drake, 2nd edn (Berkeley: University of California Press).

Gardner, M. R. (1976) 'The Unintelligibility of "Observational Equivalence"', *PSA: Proceedings of the Biennial Meeting of the Philosophy of Science Association*, 1: 104–116.

Katz, J. (1967) 'Remarks on Quine on Analyticity', *Journal of Philosophy*, 64: 36–52.

Katz, J. (1997) 'Analyticity, Necessity, and the Epistemology of Semantics', *Philosophy and Phenomenological Research*, 57(1): 1–28.

Laudan, L. and J. Leplin (1991) 'Empirical Equivalence and Underdetermination', *Journal of Philosophy*, 88(9): 449–472.

Magnus, P. D. (2005a) 'Peirce: Underdetermination, Agnosticism, and Related Mistakes', *Inquiry*, 48(1): 26–37.

Magnus, P. D. (2005b) 'Reckoning the Shape of Everything: Underdetermination and Cosmotopology', *British Journal for the Philosophy of Science*, 56(3): 541–557.

Muller, F. A. (1997) 'The Equivalence Myth of Quantum Mechanics', *Studies in the History and Philosophy of Modern Physics*, 28(1–2): 35–61, 219–247.

Newton, I. (1962) *Unpublished Scientific Papers of Isaac Newton; a Selection from the Portsmouth Collection in the University Library, Cambridge*, ed. A. R. Hall and M. B. Hall, trans. A. R. Hall and M. B. Hall (Cambridge: Cambridge University Press).

Norton, J. D. (2008) 'Must Evidence Underdetermine Theory?' in M. Carrier, D. Howard, and J. Kourany (eds), *The Challenge of the Social and the Pressure of Practice: Science and Values Revisited* (Pittsburgh: University of Pittsburgh Press), 17–44.
Quine, W. V. O. (1970) 'On the Reasons for the Indeterminacy of Translation', *Journal of Philosophy*, 67(6): 178–183.
Quine, W. V. O. (1987) 'Indeterminacy of Translation Again', *Journal of Philosophy*, 84(1): 5–10.
Reichenbach, H. (1919) Unpublished student notebook: Relativity seminar, Summer semester (University of Pittsburgh: Archives of Scientific Philosophy), HR 028-01.
Sklar, L. (1974) *Space, Time, and Spacetime* (Berkeley: University of California Press).
van Fraassen, B. C. (2002) *The Empirical Stance* (New Haven: Yale University Press).

7
Scientific Representation and the Semiotics of Pictures

Laura Perini[1]

> 'That's another thing we've learned from *your* Nation,' said Mein Herr, 'map-making. But we've carried it much further than *you*. What do you consider the *largest* map that would be really useful?'
>
> 'About six inches to the mile.'
>
> 'Only *six inches!*' exclaimed Mein Herr. 'We very soon got to six *yards* to the mile. Then we tried a *hundred* yards to the mile. And then came the grandest idea of all! We actually made a map of the country, on the scale of *a mile to the mile!*'
>
> 'Have you used it much?' I enquired.
>
> 'It has never been spread out, yet,' said Mein Herr: 'the farmers objected: they said it would cover the whole country, and shut out the sunlight! So we now use the country itself, as its own map, and I assure you it does nearly as well.'
>
> – Lewis Carroll,
> *Sylvie and Bruno Concluded*

The topic of scientific representation has received a significant amount of attention by philosophers of science. Most of the work done under this heading concerns questions about how science represents the world – focusing on the product of science, scientific knowledge. According to the current view, scientific knowledge is embodied by scientific models, so the more specific questions addressed in current work on scientific representation include questions about what models represent, how they do so, conditions for success in representing the world, and how models relate to data and to theories. Questions about the printed or uttered representations that scientists use in communication – establishing and expressing scientific knowledge – have received far less attention. In particular, little work has been done on visual representations. This is an important gap in its own right, but it also leaves the relationship between models (as embodying knowledge) and visual representations (as communication tools) quite unclear. In this chapter I will first argue for a specific approach to understanding visual representations – a semiotic analysis. That analysis offers an understanding of visual representations that, in turn, lays the groundwork for clarifying the relation between visual representations and scientific models.

1. Visual representations in scientific communication

Research publications and textbooks in the sciences typically include a variety of types of representations, such as text, numerals, and figures. There is significant variation *among* figures. Consider some common types: graphs, diagrams, maps, photographs, electron micrographs, drawings in detailed, naturalistic styles, drawings in schematic styles. These figures vary noticeably in appearance and in their subject matter. We cannot achieve a philosophical understanding of scientific reasoning until we understand the tools by which scientists present, defend, and evaluate ideas; given the prevalence of visual representations in scientific communication, this means that we cannot understand scientific reasoning until we have the capacity to explicate reasoning involving scientific visual representations, in all their diversity.[2]

Unfortunately, when it comes to visual representations, philosophers of science seem to have a blind spot. The situation is beginning to change, but the literature on scientific visual representations is remarkably small, considering the ubiquity of figures in scientific communication.[3] Why did philosophers of science largely ignore visual representations for so long, including during periods of intensive analysis of *linguistic* representation and its relation to scientific reasoning? The prominence of visual representations in science shows that they are important to *scientists*; why did that not prompt philosophical investigation into the roles images play? This behavior cannot be explained by lack of understanding of those roles; that would have prompted investigation – had it been recognized. This failure to *consider* the roles figures play can only be explained in terms of an assumption that images could not play robust cognitive roles. And that unstated assumption stems largely from unexamined, but highly influential, intuitions about pictures. In the excerpt above, Lewis Carroll brings two of those intuitions to light:

(1) An ideal visual representation *has* features matching those of its referent.
(2) An ideal visual representation represents *all* features of its referent.

These claims gain their plausibility from our experience of familiar types of pictures, like paintings rendered in naturalistic styles, and photographs. It seems that these pictures are distinctive because they replicate the visual experience we would have of their referent – copying their visible properties. And these kinds of pictures offer a wealth of detail about the visible properties of their objects. Such a picture is 'worth a thousand words'. Focus on these cases makes it seem as if the value of visual representations in general stems from the capacity to deliver lots of detailed information about the visible features of objects, via a two-dimensional array that has those same features. For such pictures, it also seems obvious that resemblance between the visible features of pictures and those of what they depict explains the content of visual representations. If resemblance determines content, it determines accuracy conditions. Prominent features of our experience with certain kinds of familiar pictures are thus taken to have normative implications in the form of effectiveness and accuracy conditions for visual representations. Thus the two basic intuitions that Carroll illuminates are closely associated with further assumptions about pictures.

The focus on naturalistic pictures, which represent visible subject matter, leads to the assumption that this is a definitive feature of visual representation:

(3) Visual representations represent visible (or at least spatial) things.

The apparent match between the visible properties of such pictures and visual experience of their referents leads to the assumption that this match is required for accuracy:

(4) The more similar a picture's visible features are to its referent, the more accurate it is.
(5) If a picture does not represent its referent as having a property that it has, the picture is inaccurate in that respect.

(6) Pictures with a naturalistic representational style, like photographs and Audubon prints, are more accurate than schematic visual representations.

The focus on the similarity between experience of pictures and visual experience of depicted scenes also affects intuitive characterization of the type of content pictures convey. We don't have experiences of resemblance between printed sentences and subject matter, so the content of visual representation is taken to be distinctively different from the content of statements. These views lead to an assumption about a further difference between visual representations and statements:

(7) Visual representations cannot bear truth.

A philosopher holding these assumptions would not expect visual representations to play any significant role in scientific reasoning. These assumptions imply that pictures cannot be true, and since two-dimensional patterns cannot have all the properties of the phenomenon at issue, they cannot be completely accurate. Visual representations would thus compromise the cogency of a scientific argument. Philosophers influenced by these intuitions would take scientists' publication of figures as a phenomenon that requires a social or psychological explanation rather than philosophical analysis. Hence the blind spot.

However, scientists do more than simply include images in their publications. They often describe figures as supporting, or as expressing, the conclusion of an article. Their evaluation of figures during peer review shows that they consider some figures essential for cogency, not a threat to it. Assumptions (1)–(7) are not only in tension with scientists' use of visual representations; they conflict with our everyday experience with maps, which are highly selective, schematic forms of representation. That is the source of Carroll's humor: he teases us with the fact that we should have known – all along – that those assumptions are false.

Intuitions about pictures thus lead us badly astray, and they are strong enough to persist in spite of conflicting personal experience. For philosophy of science, the main effect of these intuitions has been to lead philosophers to ignore pictures, but even those philosophers who hold favorable views about the capacity of visual representations to be involved in scientific reasoning are affected by them: Ronald Giere, for example, affirms (1) and (2) in his claim that 'the only perfect map of a territory would be the territory itself' (Giere, 2006: 73). He is committed to some more specific assumptions as well; his discussion of accuracy expresses assumption (4). 'Maps are of *limited accuracy* regarding included features. Relative distances on the map, for example, will not correspond exactly to relative distances on the ground' (ibid.).

This situation presents a challenge for philosophers. To understand scientific reasoning, we need to understand the various representational tools that appear in scientific communication. This requires understanding visual representations in particular, but investigating visual representations will be fruitless – or worse, misleading – unless there is a way to do so without the influence of unsupported assumptions about pictures. We need to take a more fundamental approach to understand what kind of contents visual representations convey, and how they do so. This requires, not a general theory of representation, but a more specific account of visual representations as communication tools – a theory of signs or a semiotics of visual representations, which can account for the relation between a perceptible mark on a page and the content it conveys. Ultimately, the conceptual resources of semiotics are necessary to understand the various kinds of visual representations, and clarify how each type contributes to scientific reasoning. However, semiotic analysis of pictures is also essential because it provides a way to escape the bounds of intuitions about pictures. Below I present a semiotic analysis capable of achieving both these aims. This in turn sets the stage for reflecting on the broad topic of 'representation in science', clarifying the differences between the semiotics of scientific communication and the main current framework for investigating scientific representation, in terms of scientific models.

2. A role for resemblance?

The intuitions Carroll identifies, and the set of assumptions that tend to come into play along with them, have a common root. They stem largely from deep intuitions about the importance of resemblance to visual representation. So, to develop a semiotics of scientific images that is free from the influence of unsupported assumptions, the first issue to address is the role of resemblance. The importance of resemblance to visual representation seems obvious, especially when naturalistic pictures are taken as the typical case. Most familiar pictures – like family photos – look very much like their subjects. Such pictures do not look *exactly* like their subjects, but they do provide an experience of visual resemblance that is in sharp contrast with our experience with text. We see textual representations but do not experience them as resembling their subject matter.

What is the significance of this experienced resemblance? What exactly could a resemblance relation explain?

Nelson Goodman shows that the role that resemblance can play in explaining visual representation is very limited. Resemblance is neither sufficient nor necessary for representation: the Duke of Wellington does not represent his portrait, and 'almost anything may stand for almost anything else' (Goodman, 1976: 4–5). Representation depends on something besides the intrinsic properties of the object, and besides object-referent relations like resemblance. This is obvious for linguistic representations: the shape of the written word 'cat' bears an arbitrary relation to the referent of the word, so something extrinsic to that pair of things (the visible marks and their referent) must explain why the word means what it does. This is not so obvious for pictures. A drawing of a cat looks to us like a cat. However, this relationship of visual similarity does not explain why the picture is about a cat. Just looking like a cat doesn't suffice to determine which things are representations (cats don't generally represent cats), so the status of an object as a representation depends on something besides resemblance (Files, 1996). Furthermore, the *content* of a representation depends on something besides resemblance (Perini, 2004). Because any two things resemble each other in a number of ways, resemblance between a representation and its referent is not sufficient to determine the content of a representation. A picture of a cat – a flat, rectilinear object – looks a lot more like other pictures than it does any cat. Some other factor, besides resemblance, must come into play in explaining why the picture represents a cat, and does not represent other pictures. Lopes reinforces Goodman's argument by noting that for many pictures, we can only experience a resemblance in light of understanding the picture's content – so the resemblance cannot explain the content (Lopes, 1996: 16–17).

These considerations show that communication with visual representations requires *interpretation* of the perceptible form of the image as conveying content, just as linguistic representations require interpretation of marks or sounds. However, although the fact that pictorial content depends on interpretation is an important point, it is just a starting point. Most of the explanatory work in accounting for visual representation will involve clarifying the kind of interpretation involved in pictorial comprehension. This is needed to account for the difference between linguistic and visual representation, and to identify the factors that lead viewers to see the picture in specific ways, and thus *result* in an experience of resemblance.

With the understanding that resemblance does not determine the content of visual representations, assumptions that greater resemblance between picture and referent makes for more accurate or otherwise better representation lose their intuitive appeal. These are, after all, standards that do not apply to other forms of representation. The accuracy and usefulness of textual representations does not depend on resemblance between their printed forms and their referents. A statement that fails to ascribe all its subjects properties is neither inaccurate, nor is it usually less useful in communicating reasoning, for that reason. The discussion of resemblance shows that the assumptions about pictures do not have adequate support, and since resemblance is not the explanatory key to visual representations, it shows that other resources are needed to investigate them.

3. Defining visual representations

In fact, visual representations can be characterized without appeal to resemblance. Visual representations are those in which spatial relations within individual symbols are interpreted as referring to some feature of the referent. This category captures a broad array of representations: photographs, which use spatial features of the picture to represent spatial features of the scene; graphs, which use spatial relations to represent relations among properties; a diagram of the Krebs cycle uses spatial relations to represent relations among stages in a process. Other visible features may also convey information, such as color, and these also may represent different kinds of features: in color photographs, colors convey content about the color properties of the scene depicted, but the colors of brain MRIs represent levels of activity, not color properties, of the brain. Visual representations are defined by the fact that spatial features of the symbol are *interpreted* to refer to features of the referent – they are not defined in terms of similarity with their referents. For this reason, visual representation is not restricted to visible subject matter. Assumption (3) is false. In addition, because the content of visual representations is not determined by visual resemblance, but by interpretation of their visual form, the accuracy of visual representations is not determined by visible similarity to their referents. Assumption (4) is false as well.

Seen in this light, visual representations are really a subset of representations in which some perceivable dimension is interpreted to convey content about some type of relation or dimension of variation. This can be achieved through interpretation of non-spatial dimensions: information can be conveyed through interpreting sounds, for example, using relations of time, pitch and volume, to represent relations among the referents of those features. Scientists do sometimes use such systems. For example, Geiger counters usually have an audible click system to represent radiation, as well as a dial (a visual representational readout). The *rate* of clicks increases with amount of radiation detected; the closeness of clicks in time is interpreted to convey information about the amount of radiation. However, the use of representations in which auditory or tactile dimensions of variation are interpreted to convey content are much less common than the use of visual representations, particularly in the more formal communication venues in which research is presented and evaluated. Since the semiotic analysis is in the ultimate service of understanding scientific communication, there is no loss in focusing on visual representations as defined above.[4]

Besides the spatial format of visual representations, there is another basic type of format, which allows for characterization of another broad type of representation. Linguistic representations have a serial format: the meaning of a representation is determined by a sequence of elements. Written and spoken languages, for example, depend on the sequence of sounds or characters. Perceptible dimensions, like the length of time between uttering the first and last word in a sentence, do not ordinarily contribute to the meaning of such a representation. In the typical case, it is only sequence of characters that determines meaning.

The distinction between spatial and serial formatting provides an explanation of what all visual representations have in common, and how they all differ from linguistic representations. Clarification of visual representations as involving interpreted spatial relations showed assumptions (3) and (4) are false, and also gives grounds for doubting (5) and (6). The intuition that the ideal picture matches its subject's properties suggests that visual representations that do not convey content about *all* of the subjects' properties are less accurate or less effective for this reason. The fact that the *kind* of interpretation applied to a picture's visible features varies from picture type to picture type, on the other hand, coheres with and explains the fact that only in rare cases (such as a picture that represents a picture) is there such a match. And it is entirely consistent with the selectivity of pictures, as Gombrich stresses in *Art and Illusion* (1961). Visual representations represent their subjects as having some properties, and they are non-committal about others (Lopes, 1996: 118–119). Further semiotic analysis will show that the fact that a visual representation is non-committal about some features of its subject matter does not make it inaccurate and is often an advantage in scientific communication.

4. Semiotics of visual representations: solving the diversity problem

This distinction between serial and spatial format explains the difference between visual representations and linguistic representations without appeal to resemblance. The fact that resemblance is not needed to explain the difference between visual and linguistic representations provides an alternative perspective on our intuitions about pictures, now seen to be groundless. Because spatial and other visible features can be used to represent different kinds of properties, this analysis also allows for different types of visual representations. The formatting distinction does not, however, provide the resources to explicate those different types.

This is because important differences and similarities among representations depend on more than meets the eye. The serial vs. spatial format distinction captures two different kinds of form–content relations that can be identified through consideration of individual representation–referent pairs. Stopping semiotic analysis at the level of the symbol–referent pair leaves out critical information for understanding representation. In *Languages of Art*, Goodman demonstrates the explanatory relevance of representational (or symbol) *systems* for accounting for different kinds of symbols. For Goodman, a symbol is an object that stands for, or refers to, something. Each symbol system is comprised of a set of characters (each character is an equivalence class of markings – utterance or visible mark), a set of referents, and relations correlating characters and referents (Goodman, 1976).

The need to focus on symbol systems, rather than individual symbol–referent pairs, is an important and underutilized insight. Analysis of system-level properties clarifies features of pictures that cannot be distinguished by either the appearance or cause of individual representations. Some of the system-level properties Goodman describes are surprising; he identifies different kinds of relations among the characters of a system, relations among the referents of the system, and relations between marks and the characters they instantiate. These are not the kind of features one would expect to be helpful in explicating pictures, since it seems like relations between picture form and referent are the key to what is distinctive about pictures. But again, intuitions lead us astray. Analysis at the level of symbol system provides a means to discover aspects of representations that do not depend on our starting intuitions about them – intuitions influenced by the prominence of the relations between an individual representation-bearer and its content, like resemblance, rather than the actual explanatory value of such relations.

The features of symbol systems that Goodman identifies not only provide valuable resources to clarify what makes different kinds of images different, they also yield some unexpected results. For example, this methodological approach makes it possible to define a Tarski-style definition of the concept of truth for some visual representation systems, and thus provides a reason to think that pictures can have the capacity to bear truth in the same sense in which statements do, but without reducing pictures to statements (Perini, 2005a). Assumption (7) is false. This is an important result for laying the groundwork to clarify how the representations scientists use to communicate are involved in scientific reasoning, because it shows that visual representations are not limited to playing minor roles in argumentation: they can play the critical roles of expressing or supporting the conclusion of an argument. In next three sections of the chapter I will show how application of Goodman's method addresses the diversity problem, by clarifying the different kinds of contents conveyed by different kinds of visual representations.

5. Compositional visual representations

Goodman focused on two kinds of symbol system properties: syntactic features, which have to do with aspects of the character set of a system, and semantic features, which describe relations among the referents of a system. Linguistic systems have a distinctive syntactic profile. They are syntactically articulate: all characters can be identified, and you can tell different characters apart (Goodman, 1976: 135–136). They are also syntactically disjoint: a marking instantiates at most one character (ibid.: 131–133). In such a system, if you cannot tell precisely what a marking is (like whether it's an 'a' or a 'd') it's illegible, not an additional character. These features allow for compositionality. Most familiar systems with this syntactic profile – serially formatted systems like text and numerals – also involve rules for making more complex characters out of combinations of atomic characters. The atomic characters are combined in ways such that more complex characters are also always identifiable.

The articulate and discrete syntactic profile that makes compositionality possible is one of the distinctive features of text, but it is an independent feature from the serial formatting of text. In fact, some visual representations are compositional. They derive from syntactically articulate and disjoint systems, and so have an important similarity to text. The characters of such systems are built out of combinations of

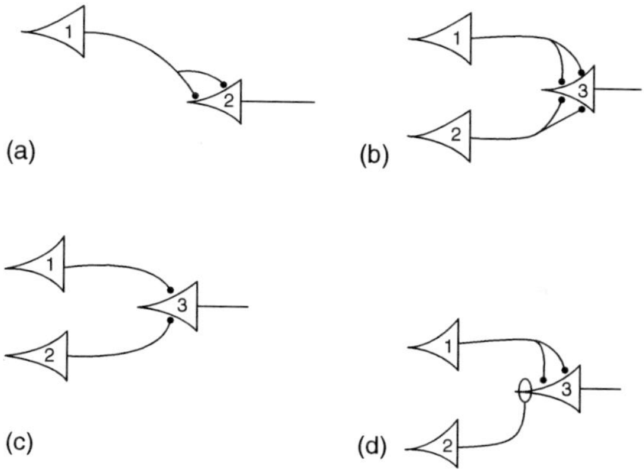

Figure 7.1 Compositional diagram: neural connections (McCulloch and Pitts, 1943: fig. 1)

discrete atomic characters, which may be combined in various ways. Examples include molecular diagrams, electrical circuit diagrams, and Figure 7.1. Spatial relations *among* atomic characters are used to represent various kinds of relations among the referents of those characters; for this reason, these figures qualify as visual representations.

This is the advantage of a more in-depth semiotics: it allows explanation of some differences among visual representations. It has already provided something unexpected, showing how some visual representations are similar to text. The method is not limited to mere classification, however; it allows for explanation of the advantages of different kinds of representations.

Figure 7.1 is a representation of neural connections. Figure 7.1(b), for example, shows neural connections in which the neuron on the right fires if and only if at least one of the two neurons on the left fires. The diagram represents these firing conditions through the use of atomic characters including the triangular shapes and lines. The triangular shapes represent cell bodies, the lines refer to cell axons; a line that ends in an open circle represents an inhibitory connection, and lines that end in a closed dot (somewhat difficult to see), end in an excitatory connection. Two excitatory connections are required for one neuron to cause another to fire. This is a very limited vocabulary, but because spatial relations among atomic characters are also meaningful, the authors can represent a variety of functional relationships. In Figure 7.1(a) one neuron alone can cause another to fire (due to two axonal excitatory connections); Figure 7.1(c) represents a case in which the neuron on the right can only fire if both those on the left do.

Abraham (2003) details the ways in which McCulloch and Pitts idealize neural *behavior*, through the assumption that neural cells have an 'all or nothing' type of behavior, either firing or not. These diagrams do not, however, involve an idealization of neural *anatomy*. Ramon y Cajal's discovery of pyramidal cells in the cerebral cortex influenced the shapes in which the cell bodies were drawn, but McCulloch and Pitts did not use the detailed information available about those cells (Abraham, 2003: 417–418). They used highly simplified shapes instead, because the characters for neurons are not meant to convey anatomical information except for the distinction between cell body and axon. Further anatomical information is simply not relevant to their aims.

This is reflected in Figure 7.1. Compositional figures represent relations among the referents of the atomic characters, but they can be non-committal about many – or all – of the intrinsic properties of those referents. The atomic characters in compositional images can function like labels, referring without representing the intrinsic properties of their referents – like the open and filled circles, which represent inhibitory and excitatory connections, respectively. The arrangement of the atomic characters conveys information about relations among the referents of atomic characters; the two-dimensional formatting makes those relations easy to comprehend and salient. Compositional figures are especially well suited to representing systems whose properties are explained in terms of relations among its component parts (Perini, 2005c). The fact that the form of atomic characters need bear no information about the referents of those characters grounds a further advantage, because irrelevant information about system components can be excluded from the diagram.

It should now be clear that compositional visual representations are not inaccurate due to lack of match between their visible properties and all their referents' properties. Accuracy depends only on whether their subjects have the properties ascribed – and one advantage of compositional diagrams is that they can refer to things with atomic characters without ascribing any intrinsic properties to those things. Information about the intrinsic properties of the referents of the atomic characters is left out, but that does not make the figure inaccurate, and it also does not make the figure imprecise, because it does not restrict the precision with which relations among those referent objects can be represented. Assumption (5) is false.

6. Pictorial visual representations

Many visual representations do not derive from compositional systems. A second type of visual representation, which I'll refer to as 'pictorial', is exemplified by photographs, images rendered in naturalistic style by hand, and images produced through image-making technologies like electron micrographs (see Figure 7.2). The systems involved in these cases have some distinctive properties. Pictorial *systems* are both syntactically and semantically dense: for any two characters, there is another character that is intermediate in form,[5] and the characters in the system map onto a dense set of referents, so there is a referent object intermediate between any two referents of the system. System density carries no implications for 'density' in terms of visual complexity in an individual symbol or 'density' in terms of information conveyed by an individual symbol in a dense system.[6] However, combination of syntactic and semantic density and visual formatting does explain the capacity of those systems to represent in great detail. Because spatial features are used to convey content, when two figures differ in form in terms of a feature in which the system is dense – no matter how slight the difference – they instantiate different characters and they differ in meaning. In this figure, changing the width of one of the dark vertical lines produces a figure that represents a different structure – a cell with a thicker axon – than the original does.

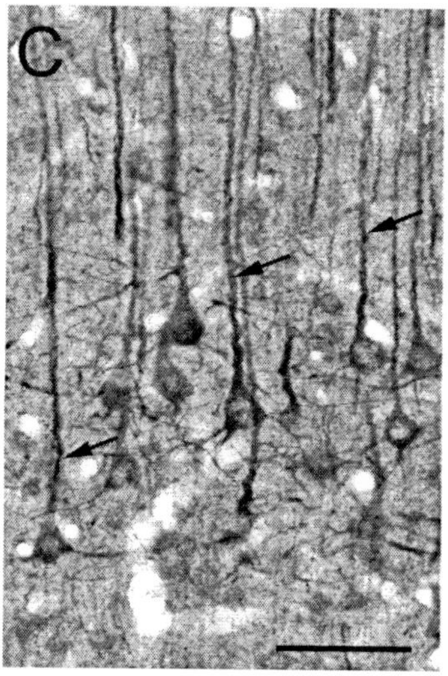

Figure 7.2 Pictorial representation: electron micrograph (Dieterich et al., 2008: fig. 2(c))

Because pictorial systems are dense, and because they use visible features to represent properties, they have the capacity to represent a dense set of very complex properties. This system-level property explains why an individual pictorial representation can convey so much detail, like the micrograph of neural cells in Figure 7.2. Pictorial visual representations provide a readily comprehensible way to represent complex content. We do not need to be able to describe its content in words in order to understand the image. The systematic relation between picture form and content thus can make it possible to communicate about features with visual representations in circumstances where there are no linguistic terms that would allow for concise, comprehensible description.

Pictorial representations include those pictures taken as paradigms of visual representation: naturalistic renderings that prompted intuitions about the importance of resemblance. The analysis of this type of visual representation has shown that it is not a match between picture and referent that explains why pictorial representation is distinctive, but the two-dimensional format and system-level syntactic and semantic properties. It is because of the syntactic and semantic density of the system – which determine the specific kind of visual representation the micrograph is – that the specific details of the shapes in the figure convey content about specific features of the biological sample. As is already clear from the discussion of compositional diagrams, in other types of visual systems the specific visible forms do *not* represent specific properties of the referent. Scientists have to apply very different kinds of interpretations to the different kinds of figures they encounter in order to comprehend them. Pictures with naturalistic style, like Figure 7.2, can be false, while diagrammatic figures can be accurate (Hall, 1996). The analysis of pictorial representation and diagrams explains why: accuracy neither depends on nor correlates with naturalistic representational style. Assumption (6) is false.

7. Schematic drawings

There is a further type of visual representation that involves neither the compositionality of diagrammatic systems nor the syntactic and semantic density of pictorial systems. These systems are not dense in the way that pictorial systems are: some visible differences do not correlate with different content. This is because it is primarily more abstract visible properties, like the contiguity of line, that are the visible features which convey content, rather than the specific visible features as is the case

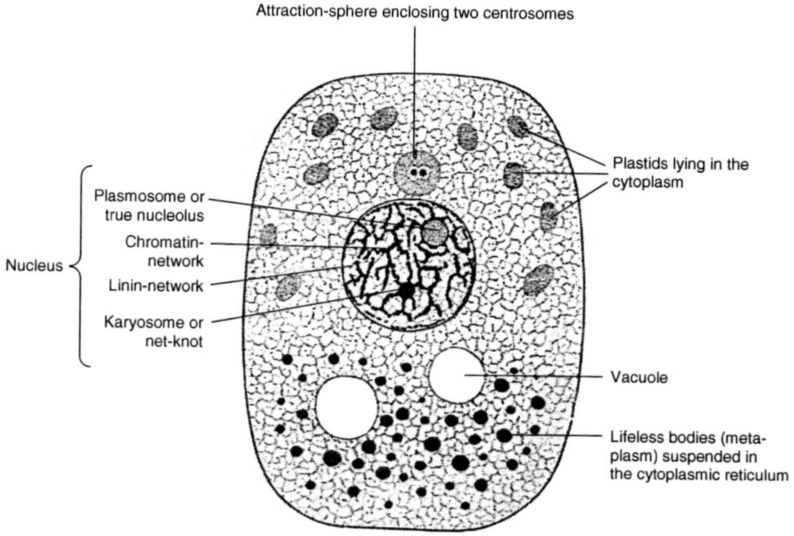

Figure 7.3 Schematic drawing: the cell wall (Wilson, 1896: fig. 2(a))

with pictorial representations. Pictorial representations convey information about very specific properties because the specific visible features of those figures are interpreted in terms of specific properties of the referent. Schematic drawings do not involve the same kind of precision. The contour depicting the cell wall in Wilson's figure, for example, is not representing the specific shape of any particular cell, or all cells (Figure 7.3). Schematic drawings also do not have the precision of compositional figures. For example, adding a few more or having a few less of the dark circular areas near the bottom would not change the content of this figure. The diagram has a specific number of spots, but it does not represent cells as having that particular number of lifeless bodies. Rather it is the more generic feature of having some spots that represents a correspondingly general property, that cells include several lifeless bodies. In these systems, relatively abstract visual features like contiguity, inside/outside, etc. are used to represent generic, rather than specific, properties.

The ability to represent generic features, without doing so via representation of the specific ways those features are instantiated, is very important in disciplines like biology, in which scientists need to communicate about features that are shared among a variety of individuals, even though those individuals vary in how those shared properties are

Table 7.1 Common types of representations in scientific communication

Symbol system features	*Serial format*	*Spatial format*
COMPOSITIONAL SYSTEM Syntactic features: characters are articulate & disjoint	Text Numerals	Compositional diagrams
PICTORIAL SYSTEM Syntactic features: characters are not articulate, character set is dense Semantic features: Referent set is dense		Electron micrographs Photographs Some line graphs Topo maps
SCHEMATIC SYSTEM Syntactic features: characters are not articulate Semantic features: referent set is not dense		Schematic drawing

instantiated. This understanding of schematic drawings shows that figures do not have to represent properties in detail to be accurate. The schematic drawing *has* a specific shape, but the specifics of that shape are not interpreted to refer to correspondingly specific properties. Again we see that assumption (6) is false.

Most pictures are non-committal about some of their subject's properties. The discussion of compositional and schematic visual representations has shown that figures are selective in two different ways, neither of which results in a loss of accuracy. Compositional figures tend to be silent about intrinsic properties of the referents of their atomic characters. Schematic drawings represent properties at a higher level of generality than do pictorial representations, leaving out information about detailed ways in which those more general properties might be instantiated.

Table 7.1 sums up the representational types discussed so far.

8. Representing in science

Semiotic analysis has had the following results. It has generated a taxonomy of scientific visual representations. This is not merely a classificatory result: the semiotic features on which that taxonomy is based do substantive conceptual work by clarifying differences in the kinds of content different types of figures convey. Semiotic analysis also showed that the intuitive assumptions about pictures are not merely groundless but false.

At this point it is also possible to clarify the relationship between analysis of visual representations scientists communicate with, and the dominant philosophical approach to scientific representation. The traditional project of understanding scientific investigation is aimed directly at understanding the product of science – scientific knowledge. For example, much recent interest in scientific laws concerns whether they are the sort of representation that tell us how things are – or whether they fail in this due either to falsity (they purport to represent how the world is, but get it wrong) or because they are not really about actual aspects of the world (Cartwright, 1983). The traditional project of studying representation in science thus concerns how scientific knowledge represents the world, in a sense that is different from asking about the communicative tools scientists use. The project of understanding visual representations begins with investigation of the meaning of those representations, and the focus is on how and why they convey the contents they do, whether or not they represent accurately or have adequate support. I believe that semiotic analysis offers a promising route to achieving the traditional, general goal of understanding scientific knowledge: once we understand the various representations presented in scientific communication, we can explicate how they function to establish and articulate scientific knowledge. My aim in the following discussion is not to forge that connection between the two projects – that is beyond the scope of a single essay – but to clarify relations between the two projects in their current state of development.

Both projects revolve around representational taxonomies. Philosophers working in the tradition of addressing how science represents the world generally *start* from a taxonomy of representations that includes one or more categories thought to play the primary representative role in science. The taxonomy is not the result of study of scientists' communication tools, however, but a product of epistemic distinctions. The logical positivists started with a distinction between theory and observation, taking theory as the primary form of scientific knowledge. Epistemically identified types of representations were conceived in terms of both logical features and the kind of content they convey: theories as sets of axioms that included theoretical terms, and observation statements as limited to a non-theoretical vocabulary. Problems arising from both the theory/observation distinction and also from the particular way theories were conceived within this framework ultimately led to the rejection of this approach. Theories were reconceived as specifying models. The dominant view of scientific representation now works with an expanded taxonomy of representations (theory, models, and data), and takes models to play the primary representational role in scientific knowledge. Clarity about both *what* models represent and *how* they do so then becomes critical to the traditional project of understanding how science represents the world. That leads philosophers studying models to questions that appear similar to those involved in studying visual representations, but it is important to not confuse them.

For one thing, there are good questions about the relationship between visual representations and models: consider a diagram of the structure of a protein, such as hemoglobin. Does the diagram represent a model, which in turn represents the protein's structure? Is the diagram itself a model? Is the proposed structure really a scientific model? We need the distinction between models and visual representations – and progress in understanding both – in order to even ask such questions.

In addition, failure to discriminate between the two different projects of understanding scientific representation leads to confusion. One source of confusion occurs when the analytical tools appropriate for one issue are applied to cases for which the other is more appropriate – analyzing figures in terms of models rather than as visual representations or analyzing models as visual representations. For example, Harris (2003) is concerned with the epistemic status of an electron micrograph. The picture is the *first* data scientists get, but it is also the result of extensive manipulation of the sample, and interaction between the prepared sample and a complex instrument, and for this reason Harris thinks that it's a mistake to think of the picture as *raw* data (Harris, 2003). Harris applies a taxonomy of representations stemming from the models literature, which leaves him with only two options: classify the micrograph as data or as a data model. Because the micrograph is the result of significant sample preparation as well as the technology involved in the electron microscope, Harris assigns it the status of data model. However, this assignment does not seem to further his goal of understanding cases in which data acquisition is the result of extensive manipulation. This is because prior discussions of data models concern cases in which there is data (for example individual values from an experiment), and starting from that data a scientist produces a data model that is a *different* representation than the original data, such as an average of those numbers. This is a 'data model', because it is a model *of* the original data. Some models of data *are* conveyed through visual representation.[7] The electron micrograph, however, is not a model of data in this sense. Harris does not show how the categorization of the micrograph as a data model provides any insight into the micrograph or its particular epistemic status that is not already captured by his discussion of why it should not be considered 'raw data'. Shoehorning the figure into the representational taxonomy of models does not further understanding of this case, because the issues Harris is interested in are not well treated by the taxonomy of data and models. These issues can, however, be addressed effectively through analysis of the figure as a visual representation. That will clarify relations among the content of the micrograph (what features it represents a particular sample as having) and the complex process involved in creating the micrograph, and the natural phenomena that the micrograph is used to study. Treating an electron micrograph as a representation in *this* sense is essential in explaining the role it plays in evidential reasoning (Perini, 2005b).

Another source of confusion stems from attempts to use visual representation to understand scientific models. This is a more serious problem than improperly analyzing particular visual representations as models, because mistaken assumptions about pictures tend to be carried over into thinking about models. This is clear in Giere's discussion of models. Although Giere has become more circumspect about characterizing pictures *as* models, he still uses maps to illuminate the nature of models *as representations*. In his discussion of maps, Giere claims maps are '*physical objects*, for example, a piece of paper with colored lines and spaces on it. It does not, therefore, make sense to ask whether a map is *true* or *false*' (Giere, 2006: 72).[8] Rejecting truth, Giere proposes that the kind of assessment appropriate to maps and models concerns the fit between the object serving as representation and an aspect of the world. Here, Giere seems to be influenced by assumptions that visual representation is essentially about matching relations, and that the accuracy of a map in particular depends on a match between map and territory. This view is extended to scientific models: whether or not a model fits some aspect of the world depends on matching it, in certain respects and to a certain degree. Semiotic analysis of visual representations has already shown that Giere's assumptions about maps are false. Obviously, appeal to false claims about pictures cannot make for a strong argument about the nature of models.

However, the apparent similarity between visual representations and models provides a strong temptation to use pictures to illuminate models. The temptation to compare visual representations and models stems in part from those problematic intuitions about the importance of resemblance to pictorial representation – pictures thus seem significantly similar to models, which are characterized by similarity or isomorphism. For that reason I will draw some of the implications of using analogical reasoning to learn about models from what we know about visual representations – now that the influence of those mistaken assumptions about pictures can be avoided. I do not present the analogy as a way of defending a view of scientific models; I have doubts about whether the two kinds of representation are appropriately similar. However, because visual representation and scientific models have been so tightly associated it is worthwhile examining what that association, if correct, suggests.

The two kinds of representation are associated because they seem to share two key features: scientific models are generally characterized both as representations and as objects that bear a similarity or isomorphism relation to what they represent. Study of visual representations has shown that, even for pictures that do look like their referents, the resemblance doesn't account for the representational status or the content of the picture. The same kinds of arguments that show that resemblance can play at most a limited explanatory role in accounting for visual representation will hold for models. This point has been recognized and has motivated some to develop accounts of scientific models that do not appeal to similarity, though with no consensus on an alternative (Frigg, 2006; Hughes, 1997; Suarez, 2003). Others continue to incorporate similarity in their accounts of models (Giere, 2006; Godfrey-Smith, 2006; Weisberg, 2007), and this is not necessarily an error. Goodman's discussion of resemblance proves only that it does not suffice to explain pictorial representation; it does not prove that resemblance is irrelevant to pictorial representation (Files, 1996). However, the fact that visual representations can be characterized without appeal to resemblance provides a reason to be cautious about what explanatory headway could be made by appealing to similarity in an account of scientific models. The upshot is that the burden is now on those appealing to similarity to clarify just what it is that similarity explains.

This challenge has not been met. For example, Giere acknowledges the fact that similarity cannot be a sufficient condition for representation. He appeals to users' intentions as an additional, and necessary, condition (Giere, 2006: 60). That allows him to avoid putting too much explanatory weight on similarity, but it does not clarify what similarity contributes to our understanding of scientific models. At a minimum, similarity must be involved in explaining the difference between the theories and models. However, for the account to serve Giere's own purposes – understanding scientific representation in the sense of how science represents the world – similarity must play a more specific role, contributing to an explanation of why it is that models represent actual aspects of the world but theories do not. Recall that analysis of visual representations showed that the feature that they all share – spatial formatting – explains how text is different from all visual representations. That fact about visual representations is the key difference between text and pictures, but it had few further implications. It does not, for example, imply that visual representations cannot bear truth, that they cannot convey propositional content, that their accuracy depends on a match between their form and their content, etc. This is because the formatting distinction is just that – it is not a fundamental difference in type of representation relation holding between an object and its referent. It does not imply the accuracy conditions expressed in assumptions (4)–(6). So by analogy, the fact that models are different from theories in virtue of having a similarity relation to the aspects of the world they represent does not imply that models should be assessed in terms of fit rather than accuracy or truth. The analogy suggests that even if scientific models are different from theories due to a similarity relation to their referents, that fact alone does not imply any fundamental difference in the nature of the representation relation involved. If theories and models are fundamentally different, it is not the similarity relation that explains this difference. Thus the appeal to similarity does not in the end do substantive explanatory work for Giere, and in particular it does not help him elaborate his position on scientific representation by clarifying how models, but not theories, represent aspects of the world.

There is another lesson to be drawn from analysis of visual representations. The visual vs. spatial formatting distinction does not explain many crucial things, such as the similarity that holds between compositional diagrams and text, and differences and similarities among various visual representations. For visual representations, it is the specifics of form–content that matters, and these specifics were only accessible through analysis at the level of symbol system. By analogy with visual representations, we cannot expect similarity to provide much insight into any particular model as a representation. The results of the semiotic analysis of pictures indicate that a much more fine-grained account of models' representational features is needed if those are to explain their epistemic role as models.

At this point it is important to stress the differences between models and visual representations. Visual representations are elements in symbol systems. Scientific models cannot be thought of as characters in symbol systems because characters are perceivable and many models, being abstract objects, are not. So the analytical tools that were so helpful in understanding visual representations are not applicable to scientific models. We are left with no clues about what sort of more fine-grained analysis would be effective for learning about models. Overall, analogy with visual representations gives a reason to think that the appeal to similarity between scientific models and their referents is an explanatory dead end.

9. Conclusion

Perhaps the most important lesson to be learned from the study of visual representations is that so many of our assumptions about pictures are flat out false. If the goal is to understand scientific reasoning, we need to understand the representations scientists use when they communicate. This requires a new approach the topic of representation in science, taking semiotics first – this is the only way to avoid the influence of those assumptions. Because scientists use printed representations to communicate about theories, models, and data, we need to study the representations used in communication in order to address questions about epistemically identified representational types, like models, and thus to pursue the project of understanding how science represents the world. The approach I've defended is essential for another reason: assumptions about pictures – like the importance of resemblance – get carried over into assumptions about models – the explanatory importance of similarity is overblown. Ultimately, by helping to escape the pernicious influence of assumptions about pictures and by getting clear on what scientists are actually talking about, semiotics is an essential part of the core project of understanding scientific representation.

Notes

1. I wish to thank Peter Kung, P. D. Magnus, and Jacob Busch for their helpful input and comments in developing this chapter.
2. Suppe (1998) argues that, since the primary means of establishing scientific knowledge occurs through published articles, not only do philosophers need to understand the nature of arguments that scientists actually present in those papers, accounts of confirmation in science must accommodate those arguments. Suppe focuses on the overall argumentative structure of scientific publications, but we cannot claim to make sense of those until we understand their components.
3. For examples of individual papers, see Wimsatt (1991), Nersessian (2002), and contributions by Giere, Ruse, Griesemer and others in Baigre (1996), and Taylor and Blum (1991). More comprehensive research efforts were taken up in several recent doctoral dissertations (Delehanty, 2005; Perini, 2002; Sargent, 1997).
4. My account excludes another type of representational system that might seem, intuitively, to qualify as 'visual'. Consider a system in which the characters are

color-chips ranging through the visible spectrum from blue to red, referring to the rational numbers 0 to 1 inclusive. This system does not involve interpretation of relations among parts of individual representations, but does interpret individual representations in such a way that there is a systematic mapping from a range of characters to a range of referents. The distinctive features of this kind of representation can be fully explicated without taking this to be a case of visual representation, because they depend on *other* properties – symbol system properties – discussed below. Visual perception plays a distinctive role in comprehending these characters, but two-dimensional formatting is not involved.

5. Because micrographs vary in two spatial dimensions as well as in tone, this is not a well-defined betweenness relation.
6. The color-chip system described in note 4 is both syntactically and semantically dense, but it is not syntactically articulate. This contrasts with the standard numerical system for representing numbers in the same range, which is syntactically articulate.
7. A smooth curve through data points is a standard example of a data model; see Giere (2006) for examples of pictorial representations functioning as data models.
8. This is a blatant non-sequitur: printed sentences are physical objects, paper with lines on it, yet printed sentences may be true or false. Once again Giere's assumptions about visual representations serve him badly.

References

Abraham, T. H. (2003) 'From Theory to Data: Representing Neurons in the 1940s', *Biology and Philosophy*, 18(3): 415–426.

Baigre, B. (ed.) (1996) *Picturing Knowledge: Historical and Philosophical Problems Concerning the Use of Art in Science* (Toronto: University of Toronto Press).

Cartwright, N. (1983) *How the Laws of Physics Lie* (Oxford: Oxford University Press).

Delehanty, M. (2005) 'Empiricism and the Epistemic Status of Imaging Technologies', unpublished doctoral dissertation, University of Pittsburgh, Pittsburgh.

Dieterich, D. C., A. Karpova, M. Mikhaylova, I. Zdobnova, I. König et al. (2008) 'Calendrin-Jacob: a Protein Liaison that Couples NMDA Receptor Signalling to the Nucleus', *PLoS Biology*, 6(2): e34. doi:10.1371/journal.pbio.0060034.

Files, C. (1996) 'Goodman's Rejection of Resemblance', *British Journal of Aesthetics*, 36(4): 398–412.

Frigg, R. (2006) 'Scientific Representation and the Semantic View of Theories', *Theoria*, 55: 49–65.

Giere, R. (2006) *Scientific Perspectivism* (Chicago: University of Chicago Press).

Godfrey-Smith, P. (2006) 'The Strategy of Model-Based Science', *Biology and Philosophy*, 21: 725–740.

Gombrich, E. H. (1961) *Art and Illusion: a Study in the Psychology of Pictorial Representation* (Princeton: Princeton University Press).

Goodman, N. (1976) *Languages of Art: an Approach to a Theory of Symbols*, 2nd edn (Indianapolis: Hackett).

Hall, B. (1996) 'The Didactic and the Elegant: Some Thoughts on Scientific and Technological Illustrations in the Middle Ages and Renaissance', in B. Baigre (ed.), *Picturing Knowledge: Historical and Philosophical Problems Concerning the Use of Art in Science* (Toronto: University of Toronto Press).

Harris, T. (2003) 'Data Models and the Acquisition and Manipulation of Data', *Philosophy of Science*, 70(5): 1508–1517.

Hughes, R. I. G. (1997) 'Models and Representation', *Philosophy of Science*, 64 (Proceedings): S325–S336.

Lopes, D. M. (1996) *Understanding Pictures* (New York: Oxford University Press).

McCulloch, W. S. and W. Pitts (1943) 'A Logical Calculus of the Ideas Immanent in Nervous Activity', *Bulletin of Mathematical Biophysics*, 5: 423–478.

Nersessian, N. J. (2002) 'Abstraction via Generic Modeling in Concept Formation in Science', *Mind and Society: a Journal of Cognitive Studies in Economics and Social Sciences*, 3: 129–154.

Perini, L. (2002) 'Visual Representation and Scientific Knowledge', unpublished doctoral dissertation, University of California, La Jolla.

Perini, L. (2004) 'Convention, Resemblance, and Isomorphism: Understanding Scientific Visual Representations', in G. Malcolm (ed.), *Multidisciplinary Approaches to Visual Representations and Interpretations* (Amsterdam: Elsevier).

Perini, L. (2005a) 'The Truth in Pictures', *Philosophy of Science*, 72: 262–285.

Perini, L. (2005b) 'Visual Representations and Confirmation', *Philosophy of Science*, 72: 913–926.

Perini, L. (2005c) 'Explanation in Two Dimensions: Diagrams and Biological Explanation', *Biology and Philosophy*, 20: 257–269.

Sargent, P. (1997) 'Imaging the Brain, Picturing the Mind: Visual Representation in the Practice of Science', unpublished doctoral dissertation, University of Minnesota, Minneapolis.

Suarez, M. (2003) 'Scientific Representation: Against Similarity and Isomorphism', *International Studies in Philosophy of Science*, 17: 225–244.

Suppe, F. (1998) 'The Structure of a Scientific Paper', *Philosophy of Science*, 65: 381–405.

Taylor, P. and A. Blum (1991) 'Pictorial Representation in Biology', *Biology and Philosophy*, 6(2).

Weisberg, M. (2007) 'Who is a Modeler?', *British Journal for the History of Science*, 58: 207–233.

Wilson, E. B. (1896) *The Cell in Development and Inheritance* (New York: Macmillan).

Wimsatt, W. C. (1991) 'Taming the Dimensions – Visualizations in Science', *PSA: Proceedings of the Biennial Meeting of the Philosophy of Science Association 1990*: 111–135.

8
Philosophy of the Environmental Sciences

Jay Odenbaugh

1. Introduction

The environmental sciences – ecology, conservation biology, climatology, and environmental economics – have largely been ignored by philosophers of science.[1] Each of these sciences occurs in a volatile juncture between science and politics. Since each offers indirect, and sometimes direct, counsel to policy-makers they are often politicized and as a result value-laden. Many contrarians argue we should not trust these theories, models, and data and the advice which depends on them in formulating environmental policy. Skepticism occurs in other sciences of course; however, the environmental sciences are subject to some of the most severe doubt (possibly in proportion to their policy relevance).[2] In this chapter, I consider three ways in which this skepticism is manifest. First, these sciences use mathematical models and simulations that are highly idealized and are coupled with very uncertain data. Why should we trust these models and simulations given that we know they are false, the data imperfect, and when policy is on the line? Second, I explore the occurrence and advocacy of moral values in the environmental sciences and consider whether it challenges the objectivity of scientists. Finally, in debates over global climate change, much is made of the apparent consensus concerning the effects of human-induced greenhouse gas emissions on average surface temperatures. However, scientific methods are structured around dissent and criticism. Is consensus-based science orthogonal or even harmful to scientific inquiry?

2. Models

Quantitative modeling is ubiquitous in the environmental sciences. One finds it in conservation biology with population viability analysis, cost-benefit analysis in environmental economics, and with general circulation models in climatology. Philosophers, historians, sociologists, and self-reflective scientists have recently expressed much interest in modeling as a research strategy (Giere, 1999; Kingsland, 1995; Oreskes, 2004; Pilkey and Pilkey-Jarvis, 2006; Sarewitz et al., 2000). In this section, I want to consider some of the philosophical issues surrounding quantitative modeling in the environmental sciences.

Traditionally, philosophers of science have focused explicitly on theories as opposed to models (however, see Campbell, 1920 and Hesse, 1966). This has changed dramatically in recent years. On a very simplistic view, scientific theories are sets of purported laws. What are scientific laws? In the philosophical literature, there is a raging debate over what if anything distinguishes laws of nature from other sorts of generalizations (van Fraassen 1989; Lange, 2000). However, if laws are at least true universal generalizations, then there are very few laws, and consequently theories, in the special sciences.[3] This is due to the fact that most empirical generalizations in the special sciences have exceptions. Most of science uses representational media epistemically less 'secure' than theories as sets of laws; they focus on models.

What are models? Scientific models are representations which are characterized in part by *idealized assumptions* – all models contain false assumptions. Thus, one hypothesis (see Odenbaugh, 2005) is that:

> Models are representations that are characterized by sets of assumptions at least one of which is idealized where an idealization is a false assumption that is useful for some purpose.

If we are to delineate which representations are scientific models from other sorts, then we must articulate the purposes for which scientists use their models. Here I am considering epistemic aims of scientists and not other sorts. There are many contenders, but here are a few possibilities. Scientists use models for anticipation of, intervention in, and understanding of the natural world; that is, to represent the world for prediction, control, and for explanation.[4] It is also important to recognize that there are a variety of objects which are used as models or representational media. For example, there are physical, scale, analogue and mathematical models, and we shall focus on the latter category.

Let's begin by considering one mathematical model used in conservation biology, a metapopulation model where a metapopulation can be thought of as a 'population of populations' which are subdivided spatially but connected by migrating organisms. In recent years, metapopulation models have been used to great effect in one of the most contentious environmental policy debates – the possible extinction of the northern spotted owl (*Strix occidentalis caurina*). The northern spotted owl is a monogamous, territorial subspecies inhabiting old-growth coniferous forests in western Washington, Oregon, and northern California. Throughout its range, it is closely associated with old-growth forest stands that provide food, thermal cover, and nest sites. Each breeding pair uses between one and three square miles of forest that is more than 250 years old. Forest fires, farm clearing, and especially timber harvest have reduced spotted owl habitats to less than 10 percent of their original area. The primary scientific question has been this: what must be done to prevent the extinction of the northern spotted owl? Biologists have attempted to answer this question with quantitative models and more specifically with detailed metapopulation models.[5] Of course, this scientific question has been posed in a very, very complicated political context involving the logging industry, environmentalism, and the law.

The theoretician Russell Lande was presented with a theoretical problem by the environmental lawyer Andrew Stahl in 1985. Stahl showed Lande the Forest Service's 'regional guide' – their plan to protect the northern spotted owl – and asked him to evaluate it. Using published and unpublished work along with the regional guide, Lande devised a metapopulation model to determine how much habitat must be minimally preserved to protect the species. To understand his model, we must define the following parameters and variable respectively. Let ε be the probability that a juvenile female inherits the territory of her mother, m be the number of territories a juvenile can disperse through before dying from predation, starvation, etc., h be the proportion of habitable territory, and p be the proportion of habitable sites that are occupied. Thus, the probability of a female *not* finding habitable territory in m trials is

$$(1-\varepsilon)(ph+1-h)^m$$

This is so because the probability of a female not finding habitable territory is determined by her not inheriting it from her mother $(1-\varepsilon)$ *and* not successfully dispersing to an unoccupied territory $(ph+1-h)^m$ (finding each territory occupied or uninhabitable in m trials). When modeling discretely growing populations – and female northern spotted owls reproduce only during a specific time in their third year – their growth rate is customarily represented as $\lambda = N_{t+1}/N_t$ which can be equal to, less than, or greater than one. A species or population is in 'demographic equilibrium' when $\lambda = 1$ and we will assume that the northern spotted owl is in a demographic equilibrium which occurs when

$$[1-(1-\varepsilon)(ph+1-h)^m]R_0' = 1$$[6]

In this equation, $R_0' = \sum_{x=0}^{\infty} l_x' f_x$ where l_x' is the probability of surviving to age x given the female has found habitable territory and f_x is the mean lifetime offspring production per female given that she finds habitable territory. Finally, we can solve the following equation:

$$\hat{p} = \begin{cases} 1 - \dfrac{1-k}{h} & \text{for } h > 1-k \\ 0 & \text{for } h \leq 1-k \end{cases}$$

Here k is the 'demographic potential' or the equilibrium occupancy of territory. What is crucial to notice is that the northern spotted owl can persist just in case $h > 1-k$. In 1987, the percentage of Douglas fir forest 200 years or older in western Washington and Oregon was about 38 percent and thus $h \approx 0.38$ and survey work suggests that about 44 percent of sites were occupied and thus $p \approx 0.44$. Finally, we can directly estimate k with the following equation,

$$k = 1 - h(1-p)$$

Thus, $k \approx 0.79$. Future forest plans suggested leaving 7–16 percent of Douglas fir forests 200 years or older. However, $1-k = 0.21 > 0.07 - 0.16$ and thus it appeared that the northern spotted owl was doomed unless forest plans were revised.

The work of biologists like Russell Lande who pioneered this metapopulation model was crucial to revising the Forest Service's habitat protection plans. Specifically, it led to the withdrawal of six old-growth forest timber sales in Oregon and Washington in 1985 and would be an important component of the 1994 adoption of the Northwest Forest Plan, which protects 8 million acres of old-growth forests (80 percent of the owl's remaining habitat, compared to 5 percent protected by the Forest Service's original owl plan). However, philosophically serious questions arise when we consider this and other models in the environmental sciences. These philosophical questions take on additional moral and political significance when economic activity, jobs, and non-human species are affected by our answers to these questions.

Our model incorporates several idealizations. For example, we have assumed that parameters ε and m are constants – they don't vary stochastically – and that the metapopulation is in a demographic equilibrium. Likewise, this model applies best when the habitats are distributed in a random manner. So, if some of the model's assumptions are demonstrably false, then why should we trust it especially when jobs and non-human species are on the line?

One suggestion is that this metapopulation model is 'approximately true' of the northern spotted owl. Unfortunately, an adequate account of approximate truth is elusive since customary accounts have fatal flaws. However, we do not have to have a fully-fledged account of approximate truth to see how this sort of defense might go. In the case of the assumption of demographic equilibrium, Lande attempts to show that it is approximately correct – that is, the rate of growth of the northern spotted owl is approximately one. He writes:

> The annual geometric growth rate of the northern spotted owl population estimated from currently available demographic data ... is $\lambda = 0.96 \pm 0.03$, which is statistically not significantly different from that for a stable population $\lambda = 1.0$ or from the approximate 1% annual decline estimated from long-term surveys. (1988: 605)

Another response to this worry is to show that a model's false assumptions do not matter with respect to the phenomena of interest (Wimsatt, 1981). If we replaced the idealizations with true or at least more realistic assumptions, the answer provided by the model to the question of interest would not be different. Remember we began with the following question, 'What must be done to prevent extinction of the northern spotted owl?' Lande was consulted to determine if the Forest Service's plan was adequate and he answered on the basis of his model that it was not. However, he notes that his model makes conservative predictions. If he had added more realistic assumptions, then the Forest Service's plan would still be inadequate and even more so. Here is what he writes:

> The model of dispersal and habitat occupancy is optimistic in several respects because of the assumptions that there is no difficulty in finding a mate, no dispersal out of regions containing suitable habitat, no demographic or environmental stochasticity (including environmental catastrophes ...), and no loss of fitness from inbreeding depression in small populations ... Violations of any of these assumptions would render population persistence more difficult ... hence this model is likely to underestimate the extinction threshold, or minimum proportion of suitable habitat in a region necessary to sustain the population. (1988: 605)[7]

It is important to be clear concerning Lande's response. If one is skeptical about mathematical modeling or scientific methodology in general, then his response would be ineffectual. However, given the target of his response, those in favor of the Forest Service's plan, it is effective.

So, when a contrarian argues that a model is untrustworthy due to its idealized assumptions, those assumptions can be 'discharged' in at least two ways. First, one can show that in fact the false assumption is approximately correct. Second, one can show that an idealization does not matter because replacing those idealizations would not change the answer it provides to policy questions of interest. Let us now turn to the issues of values in the environmental sciences and the skepticism they generate.

3. Value neutrality and advocacy

It is clear there are various moral values in the environmental sciences. Let me provide examples of how this occurs in a couple of contexts including conservation biology, climatology, and environmental economics. In standard Neyman-Pearson hypothesis testing, we have a null hypothesis H_0 and an alternate H_1.[8] Null hypotheses claim some causal factor has no effect; the alternate says that the causal factor does have an effect. Here are some illustrations. Suppose we are considering the role of hydroelectric dams in the Pacific Northwest and how they affect salmon populations. Specifically, they prevent salmon from moving easily from the Columbia River to the Pacific Ocean and back. Let our null H_0 and alternate H_1 say

H_0: Leaving irrigation water on will not decrease endangered salmon populations.
H_1: Leaving irrigation water on will decrease endangered salmon populations.

Here is another example. Suppose our null hypothesis asserts that the warming of average surface temperatures is due to chance and alternate hypotheses suggest that this warming is at least in part due to human emissions of greenhouse gases.

H_0: The warming over the twentieth century is due to chance.
H_1: The warming over the twentieth century is due to human fossil fuel emissions.

Obviously, there are two types of decisions we can make:

- We reject the null H_0 given that it is true or false.
- We accept the null H_0 given that it is true or false.

A *Type I error* occurs when we reject the null when it is true and a *Type II error* occurs when we accept the null when it is false. Given that our environmental data are samples from 'populations' under investigation and how difficult that data can be to collect, these errors are always possible. Customarily, scientists minimize Type I errors; they reject H_0 just in case Pr(Rejecting H_0/H_0 is true) < 0.05.[9] We could choose to minimize the risk of Type II errors, but for mathematical reasons we cannot minimize both types of errors. This choice, however, implicitly assumes that the consequences of a Type I error are greater than those of a Type II error which, given how declining salmon populations, climate change, and other environmental issues affect human well-being economically, ecologically, and aesthetically, certainly cannot be assumed. This is tantamount to assuming that environmental risks are less important and places a burden of proof on environmentalists to demonstrate a causal connection between, say, dams and salmon decline or fossil fuel use and global climate change. Regardless of one's view of which error probability is best to minimize, in the environmental sciences, facts and moral values are often mixed even in statistical hypothesis testing.

Neyman-Pearson statistical hypothesis testing is a very common methodology taught in statistics courses and used by practicing environmental scientists. However, some philosophers and statisticians reject it for an alternative, Bayesianism. Bayesians argue that we should not accept or reject hypotheses; rather, we consider how evidence affects their probabilities.[10] Nevertheless, whether similar problems affect a Bayesian inferential framework is an open question.

Here is another example of how the environmental sciences are value-laden coming from environmental economics. Recently, Sir Nicholas Stern, who is a top economic adviser to the British government, came to the United States to lecture on global climate change. He led a commission which released *The Economics of Climate Change: the Stern Review* (2007) which called for 'urgent action' on climate change. After the publication of the report, he went to Yale University to debate his proposal which essentially is this: Given the fact that we are warming the Earth in an unprecedented manner and that this is likely to lead to severe effects like droughts, species extinction, rising sea levels, and melting sea ice, we must take serious action such as carbon trading or taxes. Stern was rebutted in debate by economists William Nordhaus and Robert Mendelsohn. The main argument they raised concerned discount rates and future generations. Customary economic practice has it that future benefits and costs are of less value than present benefits and costs and usually one uses an equation like the following one to compute present value from future benefits/costs:

$$PV = \sum_t FV_t/(1 = r)^t$$

Here PV is present value, FV_t is future value at t, and r is a discount rate. So, $200 dollars twelve years from now at a 6 percent discount rate is worth $100 now. Stern's report essentially claims that present benefits/costs are of much the same value as future benefits/costs (adjusted for inflation) and Nordhaus and Mendelsohn use a much higher discount rate. From a moral point of view, reasonable people believe that it is morally wrong to discriminate with respect to race, gender, and so on. Similarly, one can argue there is no justifiable reason discriminating against individuals for living at a time later (or earlier for that matter) than we do.

There are several arguments offered in favor of discounting future value. First, individuals have 'positive pure time preferences' meaning that they prefer present benefits more than future benefits. However, there is a crucial difference in what people *do* prefer and what they *ought* to prefer. Second, economists argue that present benefits are more 'certain' than future benefits. Still, it is not necessarily true than we can be more confident of present benefits as opposed to future benefits since this depends on the content of the preferences. Third, some economists suggest that we must compare future benefits against resources being invested now at current rates of interest. One reply to this is that this argument holds only if alternative goods are substitutable and there is reason to believe that certain goods necessary for human well-being – clean air, clean water, photosynthetic organisms producing CO_2, etc. – are not substitutable with future economic benefits since we would not be present for those substitutes. Regardless of one's view about the correct value of the discount rate, the value assigned to the discount rate is thus a normative, moral issue.[11]

So, we have ample reasons to reject the claim that the environmental sciences are 'value-neutral'. The values are in many instances moral (Odenbaugh, 2003). However, there are two questions we should ask. First, what values are at stake in environmental science? Second, how should they affect our scientific decision-making? To help us understand the various values at stake it is useful to consider the work of Helen Longino (1992). Longino carefully contrasts *constitutive* and *contextual values* in science. Constitutive values are values that make science the practice that it is. As examples, values like simplicity, empirical accuracy, unification, and fruitfulness are often used in choosing hypotheses. These features are of value because of the aims which they serve in scientific inquiry and they help define what science is. Contextual values are values that arise in contexts in which science is done. Specifically, they derive from individual scientists, science as a collective, and society at large in which science is done. Personal contextual values include those economic, aesthetic, and moral values informing the work that individuals select, the analysis they do, and what kind of person they are. Collective contextual values include the practical aims of these programs both in terms of the implications of their research for future intellectual activities but also for others' well-being. For example, the investigation of biodiversity has these broad values in the sense that it may be bought or sold in markets, can improve our lives in non-material ways, and can be valued for its own sake.

It is more controversial as to how these values should affect scientific decision-making. The controversy depends on what the values are, who holds them, and what justification they have for holding them. For scientific work to affect policy-making, the science must appear to be sound; that is, not appear to be 'junk science'. If a scientist advocates science based on *uniquely personal contextual values*, then the work can appear subjective and thus irresponsible. For example, if one advocates that dams should be prohibited because of the intrinsic moral value of salmon populations and this value is not widely shared, then one risks critics challenging the work on the basis of those and not the work's epistemic quality. If a scientist advocates based on *collective contextual values shared by other scientists*, then the work can appear more objective since those values are not idiosyncratic. Finally, if one advocates based on *collective contextual values shared by other scientists and non-scientists*, then the work will appear even more objective since the values are not idiosyncratic to scientists but shared by society. For example, if general circulation models suggest that global mean temperature will increase significantly due to greenhouse gas emissions and this would have broad negative effects on human well-being, then advocating curbing emissions would appear more objective.[12] As another example, biodiversity provides all of us with essential ecosystem services like purification of water and soil, generation and renewal of soil, pollination of crops and plants, control of agricultural pests, and partial stabilization of climate. Society as a whole values these services (and should), and they should be promoted.

In sum, the environmental sciences are not value-neutral. Still, this does not imply that every value should enter scientific decision-making in the same manner. Scientists should be aware of whether those contextual values are personal or collective and how deeply they are shared. They risk the loss of credibility when those values appear to be idiosyncratic to ecologists or cannot be justified on the basis of considerations of the well-being of others. This is crucial because we live at a time where scientific credibility is under heavy scrutiny and the opinions of scientists are challenged due to possible political biases and the clear biases of those who evaluate the science for political purposes. If the public believes that the environmental sciences should be value-free (contrary to what is the case), then scientists must engage in activism or advocacy with care.[13]

4. Consensus science

Many contrarians, and others of course, make a rather obvious claim:

> (C) Scientific consensus about *p* does not make or guarantee that *p* is true.

Why accept (C)? We all know of episodes in the history of science in which there was (near) unanimous assent to some proposition but that proposition was later shown to be false. For example, scientists believed that the Earth was at the center of our solar system and that species were not related by common descent. Examples such as these are sufficient to show that propositions are not made true by unanimous consent.[14] Thus, many contrarians assert, to concern ourselves with scientific consensus is simply to let extra-scientific considerations run science. More specifically, we are allowing left-leaning political concerns to dictate the content of scientific claims. Likewise, by focusing on consensus we are reducing dissent and that undercuts the scientific process.[15]

One can argue that by focusing on consensus we do damage to scientific inquiry. Philosophers of science have recognized that even if scientists are primarily motivated by considerations other than truth – interest in peer recognition say – this can be conducive to scientific progress. For example, suppose a scientist *A* is very much wedded to a particular hypothesis. It is in his interest for it to be highly confirmed.

However, scientist *B* accepts a contrary hypothesis and thus it is in her interest to challenge *A*'s claims. Finally, allies of *A* will use *A*'s work in their own investigations; hence, it is in their interest to make sure that *A*'s hypothesis is genuinely highly confirmed and likewise for *B*. These social processes create a climate of critical engagement and thus dissent is a crucial element in weeding out reliable from unreliable work, significant truths from falsehoods. These mechanisms at the individual level have 'personal' ingredients that are harnessed socially to produce a mechanism for scientific progress. Hence, dissent is crucial for the discovery of significant truths and this may be so even if individual scientists are not particularly interested in truth *per se*. So, why should consensus matter with regard to issues of anthropogenic climate change, biodiversity loss, or evolution where it is often discussed?

To zero in on the debate, let's use the following claim as an example:

> (GW) Average surface temperatures are increasing in part because of human greenhouse gas emissions.

Naomi Oreskes (2004) has argued forcefully that there is close to complete agreement amongst professional climate scientists on (GW)'s truth. She and her graduate students surveyed through the Web of Science over 928 abstracts of articles published in peer-reviewed journals with the search term 'global climate change'. She found no essay that disagreed with the claim 'Global climate change is occurring, and human activities are at least part of the reason why.' So what? We know that there are dissenters with regard to (GW). Why does their opinion not receive equal or even greater weight?

One common argument given has been expressed in Ross Gelbspan's book *The Heat is On* (1997). Gelbspan documents how contrarian scientists have been funded by the oil and gas industry and many in this industry have generally been skeptical of (GW).[16] The conclusion he draws is that these contrarians deny (GW) for financial gain. Let's consider one example. Patrick Michaels is a Professor of Environmental Science at the University of Virginia. He is affiliated with the George C. Marshall Institute and the Cato Institute both of which are conservative think tanks. Gelbspan and others claim that he has received more than $115,000 from coal and energy interests. A quarterly publication, *World Climate Review*, which Michaels founded, was funded by the contrarian group Western Fuels. Finally, he was paid $100,000 by the electric utility Intermountain Rural Electric Association which also is contrarian in nature. This is a suggestive argument; however, it is a circumstantial ad hominem. That is, we are conflating the denial of (GW) and Michaels' associations. To correct the 'Gelbspan argument' we would need to show that Michaels denies (GW) *because of* the money he has received which requires more evidence than Gelbspan has supplied.[17]

While I wholeheartedly agree that dissent is absolutely essential to scientific advances, in other contexts, dissent of the minority stands in the way. Specifically, dissent is essential to scientific inquiry; however, when scientific hypotheses are brought to bear on policy matters, consensus is incredibly important. This is because we are now considering the beliefs of policy-makers who do not have the relevant expertise in the matters of interest. Suppose you are not a climatologist but a policy-maker; should you believe (GW)? Well, if the majority of climatologists do believe it and you are in no position to seriously evaluate the evidence on your own, then it seems that you should. Your evidence will not directly involve tree rings, glacial retreat, ice cores, satellite measurements, etc. since you don't understand those topics. Rather, you must determine who is a reasonable authority on the issue and form your beliefs in accordance with their opinion.[18] As a way of conceptualizing the point, consider the following simple model. Suppose that

- the correctness of a expert's opinion is probabilistically independent of other experts, and
- the probability that expert i is correct about a proposition q is p_i.

Hence, the probability that expert i is wrong about q is $(1 - p_i)$. If the number of experts that agree with the proposition is n, then the probability that they are all correct is $\prod_{i-1}^{n} p_i$. Similarly, the probability that they are all wrong is $\prod_{i-1}^{n} (1 - p_i)$. Hence, as n increases, the probability that they are all wrong gets smaller and smaller.[19] It is unreasonable to think that each expert i's judgement is independent of each other statistically speaking. Hence, more realistically, we might make the model more complex by using the appropriate conditional probabilities or suppose that research groups are statistically independent of one another though individuals in those groups are not (i now represents a token research group). Still, the points are these: first, if the majority of scientists accept (GW), then as that number grows the probability that they are wrong shrinks, and second, one should form one's degree of belief in (GW) in accordance with this majority opinion.[20]

The most reasonable method for doing this in a scientific context is to determine either (a) who are the relevant peer-reviewed scientists who are competent with regard to (GW) or (b) who are the scientific bodies composed of such individuals that offer summary statements about (GW)'s truth. The Intergovernmental Panel on Climate Change (IPCC) is just such an organization. The IPCC is one of the largest science-related projects ever created. It does not conduct research of its own; rather, it evaluates the work of scientists around the world and then synthesizes this work for policy-makers in a report and a summary. There are three working groups. The first works on the physical science behind climate change. The second considers the impacts, adaptation, and vulnerability due to climate change. The third group looks at mitigation or how we can reduce the climate change which is occurring and will occur. Each IPCC assessment involves between 100 and 200 researchers who are nominated by their own government or non-governmental organizations. There are two scientists in charge of each working group – one from a developed and one from a developing country. These researchers produce draft reports which are then reviewed by several hundred expert reviewers. Finally, these documents are further examined by reviewers within each government and then accepted at the plenary meeting. It should also be noted that the IPCC is not the only organization to have stated that humans are having an effect on our climate through greenhouse gas emissions. The American Academy of Science, the American Meteorological Association, the American Geophysical Union, and the American Association for the Advancement of Science have all issued similar statements.

So, here is one way consensus matters – consensus among independent inquirers is indirect evidence for a proposition's truth when you lack the relevant expertise. Scientific consensus contrary to contrarians does not threaten the process of science. Dissent is crucial for as we say 'normal science'. Contrarians should continue their criticism and battle it out in peer-reviewed journals (which they generally do not do). However, when it comes to non-expert opinion and specifically environmental policy, consensus matters a lot. This point simply pivots on drawing a distinction between what happens *inside* science and what happens *outside* of science as an institution. Though they are often inextricably connected, they are conceptually distinct.

5. Conclusion

The environmental sciences find themselves inextricably linked to both science and politics. In this chapter, I have considered three philosophical issues in the environmental sciences which arise from and are exacerbated by being in this interdisciplinary juncture. First, we have considered the role of quantitative modeling in the environmental sciences and how scientists deal with idealizations and uncertainty. Second, we found that the environmental sciences are laden with moral values though there is reason to be skeptical of the advocacy by scientists of those values. Third, it was argued that consensus science is crucial when it comes to informing policy-makers; however, this is completely consistent with the importance of dissent in scientific inquiry.

Notes

1. However, there have been some important recent works: for example, Cooper (2003); Parker (2006); Sagoff (2004); Sarkar (2005); and Shrader-Frechette and McCoy (1993).
2. A much-discussed instance of this contrarianism is Björn Lomborg's *The Skeptical Environmentalist* (2001). In this book, he criticizes the theories, models, and data of ecology, economics, and climatology.
3. This is not to say that there are no candidates for 'lawhood' in the environmental sciences. Insofar as climatology imports basic physics and physics has laws, then climatology has laws. Likewise, some argue (Colyvan and Ginsberg, 2006; Lange, 2000; and Mikkelson, 2003) that the species area relation $S = cA^z$ is a law in ecology where S is number of species, A is area, and c and z are parameters. It should also be noted that philosophers like Nancy Cartwright (1983) have argued that even basic physics lacks true universal generalizations. Her views have changed over the years, but most recently (1999) she argues that laws are true of 'nomological machines' – controlled experimental setups.
4. Whether or not there are uniquely scientific purposes can be debated. The sort of pragmatism articulated here does not see science as distinct from common sense or ordinary aims, so that issue would not be of great significance.
5. In what follows, I present the demography of northern spotted owls and modeling work of biologist Russell Lande leaning heavily on his (1988) essay. For a popular treatment of the conflict between logging and environmentalists along with parts of the story below see Durbin (1998). Likewise, for a general discussion of the natural history and science of temperate rainforests in the Pacific Northwest see Norse and Raven (1988).
6. Note in this equation we have $[1 - (1 - \varepsilon)(ph + 1 - h)^m]R_0' = 1$ which is the probability of a female successfully finding habitable territory in m trails and R_0' is a measure of fecundity. Multiplied together they give us the population's growth rate. We are assuming the metapopulation is in demographic equilibrium which means that the ratio of the population sizes N_{t+1}/N_t is equal to one and thus is unchanging.
7. In Lande's (1987) mathematical paper, 'Extinction Thresholds in Demographic Models of Territorial Populations', he examines more realistic models in addition to the basic model presented here. In effect, he explores how changing the model assumptions affects the model's dynamics.
8. For a good discussion of the value-ladenness of Neyman-Pearson testing in ecology, see Shrader-Frechette and McCoy (1993).

9. Of course, there is nothing sacrosanct about a significance level of 0.05. Sometimes we use 0.01 instead. However, significance levels themselves presuppose judgements about what standard of evidence is good enough and that depends on our aims and interests which can be ethical. Likewise, the quality of evidence we have in conservation biology and climatology is rarely this good.
10. Even if Bayesians argue that we should not accept hypotheses, we still must decide which to use for certain purposes and they need a criteria for choosing which hypothesis to use for particular purposes on particular occasions.
11. For an excellent discussion of the moral and economic issues surrounding the discount rate, see Broome (1992) and for a more elementary discussion see O'Neill et al. (2008).
12. I am assuming that generally people have moral concern for the well-being of others. Also, the more generally a value is shared, the less controversial and more objective it will appear.
13. It should also be noted that science education should challenge the myth of value neutrality. One way of improving the public's view of science is by the recognition that it is in the service of the common good. How exactly this is to be done is complex of course.
14. A technical aside: there are propositions which are made true by universal assent; namely those whose truth conditions concern universal assent. However, those very propositions are not made true by universal assent *to them*. Also, on one type of pragmatist theory of truth, a proposition is true just in case it would be agreed with by fully informed and fully rational inquirers who had an indefinite amount of time to investigate them. Here I am assuming such a theory is false.
15. For a popular argument for these sorts of claims see Horner (2007), especially his chapter 5. Ironically, the claims for much of the basis of fiction writer Michael Crighton's critique of climate change science in his novel *State of Fear*.
16. Recently, due to the work of Joseph Romm, the term 'contrarian' has been sometimes replaced with the terms 'denier' and 'delayer' with regard to global climate change. A denier denies the truth or justification for (GW). A delayer accepts (GW) but claims that we are impotent to do anything about it, it would be too expensive, or there are more efficient ways of using our GNPs. Examples of these two positions would be Patrick Michaels and Bjorn Lomborg. A contrarian is typically a denier but some are delayers.
17. It is also important to note that the funding of those who agree with (GW) does not come from a similarly but oppositely biased group since much of scientists' funding comes from agencies like the NSF. Thus, I do think that the correlation between anti-(GW) organizations and the opinions of the contrarians is *prima facie* evidence of bias.
18. I should note that this point is not unique to the environmental sciences, far from it. For example, the epistemic justification for many of our beliefs depends on the reliability of authorities. For example, most of us could not prove the fundamental theorem of calculus which shows how the differentiation and integration are related. However, in our mathematical work, we presuppose that it is true and can be shown to be so.

19. Incidentally, this is an instance of Condorcet's jury theorem provided we assume each expert *i*'s 'vote' is statistically independent of the others and each $p_i > ½$. When each $p_i \leq ½$ then a correct vote is maximized by having one and only one voter.
20. For example, one might propose the following: one's degree of belief *p* in a proposition *q* should be the average value of the majority expert's degree of belief in *q*.

References

Broome, J. (1992) *Counting the Cost of Global Warming* (Cambridge: White Horse Press).
Campbell, N. (1920) *Physics: the Elements* (Cambridge: Cambridge University Press).
Cartwright, N. (1983) *How the Laws of Physics Lie* (Oxford: Oxford University Press).
Cartwright, N. (1999) *The Dappled World: a Study of the Boundaries of Science* (Cambridge: Cambridge University Press).
Colyvan, M. and L. Ginsberg (2006) *Ecological Orbits: How Planets Move and Populations Grow* (Oxford: Oxford University Press).
Cooper, G. (2003) *The Science of the Struggle for Existence: On the Foundations of Ecology* (Cambridge: Cambridge University Press).
Durbin, K. (1998) *Tree Huggers: Victory, Defeat & Renewal in the Northwest Ancient Forest Campaign* (Seattle: Mountaineers Books).
Gelbspan, R. (1997) *The Heat is On: High Stakes Battle Over Earth's Climate* (New York: HarperCollins).
Giere, R. (1999) *Science without Laws* (Chicago: University of Chicago Press).
Hesse, M. (1966) *Models and Analogies in Science* (Oxford: Oxford University Press).
Horner, C. (2007) *A Politically Incorrect Guide to Global Warming (and Environmentalism)* (Washington, DC: Regnery Publishing, Inc).
Kingsland, S. (1995) *Modeling Nature* (Chicago: University of Chicago Press).
Lande, R. (1987) 'Extinction Thresholds in Demographic Models of Territorial Populations', *American Naturalist*, 130: 624–635.
Lande, R. (1988) 'Demographic Models of the Northern Spotted Owl (*Strix accidentialis caurina*)', *Oecologia*, 75: 601–608.
Lange, M. (2000) *Natural Laws in Scientific Practice* (Oxford: Oxford University Press).
Lomborg, B. (2001) *The Skeptical Environmentalist* (Cambridge: Cambridge University Press).
Longino, H. (1992) *Science as Social Knowledge* (Princeton: Princeton University Press).
Mikkelson, G. (2003) 'Ecological Kinds and Ecological Laws', *Philosophy of Science*, 70: 1390–1400.
Norse, E. and P. Raven (1988) *Ancient Forests of the Northwest* (Washington, DC: Island Press).
Odenbaugh, J. (2003) 'Value, Advocacy, and Conservation Biology', *Environmental Values*, 12: 55–69.

Odenbaugh, J. (2005) 'Idealized, Inaccurate but Successful: a Pragmatic Approach to Evaluating Models in Theoretical Ecology', *Biology and Philosophy*, 20: 231–255.

O'Neill, J., A. Holland, and A. Light (2008) *Environmental Values* (New York: Routledge).

Oreskes, N. (2004) 'The Scientific Consensus on Climate Change', *Science*, 306: 1686.

Otto, S. and T. Day (2007) *A Biologist's Guide to Mathematical Modeling in Ecology and Evolution* (Princeton: Princeton University Press).

Parker, W. (2006) 'Understanding Model Pluralism in Climate Science', *Foundations of Science*, 11(4): 349–368.

Pilkey, O. and L. Pilkey-Jarvis (2006) *Useless Arithemetic: Why Environmental Scientists Can't Predict the Future* (New York: Columbia University Press).

Sagoff, M. (2004) *Price, Principle, and the Environment* (Cambridge: Cambridge University Press).

Sarkar, S. (2005) *Biodiversity and Environmental Philosophy* (Cambridge: Cambridge University Press).

Sarewitz, D., R. Pielke Jr., and R. Byerly Jr. (2000) *Prediction: Science, Decision-Making, and the Future of Nature* (Washington, DC: Island Press).

Shrader-Frechette, K. and E. D. McCoy (1993) *Method in Ecology: Strategies for Conservation* (Cambridge: Cambridge University Press).

Stern, N. (2007) *The Economics of Climate Change: the Stern Review* (Cambridge: Cambridge University Press).

van Fraassen, B. (1989) *Laws and Symmetries* (Oxford: Oxford University Press).

Wimsatt, W. (1981) 'Robustness, Reliability, and Overdetermination', in M. Brewer and B. Collins (eds), *Scientific Inquiry and the Social Sciences* (San Francisco: Jossey-Bass), 124–163.

9

Value Judgements and the Estimation of Uncertainty in Climate Modeling

*Justin Biddle and Eric Winsberg**

1. Introduction

It is uncontroversial that scientific research, especially scientific research that has important public policy implications, involves value judgements. This is nowhere more evident than in research on the impact of carbon emissions upon global climate. The looming prospect of severe, anthropogenic climate change is forcing us to make difficult moral and political decisions. What kind of action should be taken to curb global climate change? How much should we value our own safety, comforts, and economic opportunities in comparison to those of future generations? How much scientific evidence do we need before taking action? Should this action be voluntary or legally mandated? The importance of these questions for our future is well known, although there is still much disagreement over the appropriate way to answer them.

Yet, there are some questions regarding the role of value judgements in climate research that are not so well known, even within the climate modeling community. In particular, while it is clear that value judgements play an important role in deciding how a given area of research should lead us to *act*, it is less clear whether such value judgements should play a role in deciding what to *believe*. In other words, while value judgements clearly play a legitimate role in the realm of *practice*, do they also play a legitimate role in the realm of *theory*?

Philosophers of science are increasingly concerned with issues such as this, although, not surprisingly, different philosophers treat these issues in very different ways. The traditional view maintains, first of all, that the following two distinctions can be drawn clearly: the distinction between theory and practice and the distinction between so-called 'epistemic' and 'non-epistemic' values. It then maintains that only 'epistemic' values play a legitimate role within the realm of theory; 'non-epistemic' values can, and should be, confined to the realm of practice.[1] According to one influential interpretation, 'epistemic values' – which include such values as simplicity, explanatory power, internal consistency, and consistency with surrounding theories – are values that are truth-conducive, in the sense that if theory T_1 exhibits a given 'epistemic' value and theory T_2 does not, then T_1 is, *ceteris paribus*, more likely than T_2 to be true; non-epistemic values, conversely, are not truth-conducive, and thus should be excluded from the realm of theory (McMullin, 1983).[2]

At various points during the last century, this traditional view has been called into question. In the early and mid-twentieth century, scholars such as C. West Churchman (1948, 1956), John Dewey (1929), Philip Frank (1954), Otto Neurath (1913), and Richard Rudner (1953) argued that values traditionally thought of as non-epistemic – including, in some cases, ethical and political values – play an inevitable role in the epistemic evaluation of research. Increasing numbers of contemporary philosophers of science are arguing for this same conclusion.[3] Despite the work of those who question this ideal of value-neutrality, however, the traditional view remains the dominant one.

In this chapter, we investigate the question of whether it is reasonable to expect climate modelers to exclude 'non-epistemic' values from the 'internal' aspects of their research – i.e. from the realm of theory. We argue that it is not. We begin our argument, in section 2, by discussing one of the most influential arguments for the claim that ethical values play an ineliminable role in the evaluation of scientific research, namely Richard Rudner's argument from inductive risk (Rudner, 1953); we follow this with a discussion of one of the most influential objections to this view, due to Richard Jeffrey (1956). Jeffrey's argument is the *locus classicus* of the view that one can distinguish clearly between the epistemic and the practical appraisal of theories and that the epistemic appraisal of research can and should be neutral with respect to 'non-epistemic' value judgements. Jeffrey's argument, moreover, is followed by many contemporary philosophers of science.[4]

We believe that an investigation of climate modeling is a particularly fruitful way to test the feasibility of Jeffrey's argument, because climate modelers, together with statisticians, attempt to do something very similar to what Jeffrey recommends – namely, to assign probabilities to hypotheses concerning the effects of carbon emissions upon global climate change in a manner that is free from 'non-epistemic' considerations. In section 3, we distinguish between the various sources of uncertainty in the development of climate models, and we discuss in general terms the ways in which climate scientists attempt to quantify these uncertainties. In sections 4 and 5, we discuss in detail two of these sources of uncertainty – structural model uncertainty and parameter uncertainty, respectively – and we argue that it is unreasonable to expect climate scientists to provide estimations of these uncertainties in a manner that is free from practical, or 'non-epistemic', considerations.[5]

Before we proceed to our discussion of Rudner and Jeffrey, there is one preliminary point that needs to be made right from the start. There is a broad consensus within the scientific community that carbon emissions are causally related to global climate change; we are convinced that this broad consensus is justified, and it is in no way our aim to call this consensus into question. We believe this consensus has been reached objectively. As will become apparent, we do not believe that the influence of 'non-epistemic' considerations on the estimation of uncertainties implies that climate models are unreliable. What it does imply is that, *pace* Jeffrey, one cannot distinguish sharply between the realms of value-neutral theory and value-laden practice. Moreover, it highlights the fact that more attention needs to be paid to the areas within climate modeling in which values play an ineliminable role, to the kinds of values or practical considerations that play a role, and to the effect that these values have upon the overall performance of our models. None of this, however, should be taken as evidence for skepticism about climate change. With this preliminary note aside, we can now proceed to our discussion of Rudner and Jeffrey.

2. Rudner and Jeffrey on the role of ethical values in science

In an essay entitled 'The Scientist *Qua* Scientist Makes Value-Judgments' Rudner argues that ethical values play an inevitable role in the epistemic appraisal of hypotheses. His argument proceeds as follows:

1. The scientist *qua* scientist accepts or rejects hypotheses.
2. No scientific hypothesis is ever confirmed with certainty. In accepting or rejecting a hypothesis, there is always the possibility of being wrong.
3. The decision to accept or reject a hypothesis depends upon whether the evidence is sufficiently strong.
4. Whether the evidence is *sufficiently* strong is 'a function of the *importance*, in a typically ethical sense, of making a mistake in accepting or rejecting the hypothesis' (Rudner, 1953: 2, his emphasis).
5. Therefore, the scientist *qua* scientist makes value judgements.

To illustrate this argument, Rudner considers the evaluation of the following two hypotheses: (a) a given drug, which we know contains a toxic ingredient that is lethal to human beings, is safe for human consumption, because it does not contain this ingredient in dangerous quantities; (b) a given batch of machine stamped belt buckles is not defective. Rudner argues that we require a much higher standard for accepting the former hypothesis than the latter, because the moral consequences of wrongly accepting the first hypothesis are much more serious than the second. 'How sure we need to be before we accept a hypothesis will depend on how serious a mistake would be' (Rudner, 1953: 2). Given that the degree of confirmation that we require in order to accept (or reject) a given hypothesis depends upon an evaluative judgement regarding potential ethical consequences, ethical considerations play an inevitable role in the appraisal of hypotheses. It should be noted that, in drawing this conclusion, Rudner is maintaining that value judgements, *even in the ideal*, play a role in the appraisal of hypotheses. It is not only the scientist *qua* human being – i.e. *qua* individual who is invariably influenced by the prejudices of her time – who makes value judgements; it is also the scientist *qua* scientist.

While Rudner does not spend much time examining the implications of this conclusion for a broader theory of science, he does discuss briefly its implications for the notion of scientific objectivity. According to a traditional interpretation, scientific objectivity requires that hypotheses be evaluated in a value-neutral fashion – or at least in a fashion that is neutral with respect to ethical values. Rudner does not deny that objectivity is and should be an ideal for scientific inquiry, but he argues that the necessarily value-laden character of hypothesis appraisal implies that this traditional interpretation of objectivity is misguided: 'What seems called for . . . is nothing less than a radical reworking of the ideal of scientific objectivity' (Rudner, 1953: 6). Rudner does not undertake the task of reworking this ideal; he does, however, argue that objectivity in science 'lies at least in becoming precise about what value judgments are being and might have been made in a given inquiry' (Rudner, 1953: 6). In other words, if value judgements do play an inevitable role in scientific research, it would be in the interests of objectivity to know precisely which value judgements are playing a role, where they are playing a role, and the effect that these judgements have upon the research in question.[6]

Many, however, deny that value judgements of an ethical – or more broadly, 'non-epistemic' – character play an inevitable role in the epistemic appraisal of hypotheses; one of the clearest arguments for this view was put forward by Jeffrey, in direct response to Rudner (Jeffrey,

1956). Jeffrey argues that the scientist *qua* scientist does not accept or reject hypotheses but merely assigns probabilities to them: 'the scientist's proper role is to provide the rational agents in the society which he represents with probabilities for the hypotheses which on the other account he simply accepts or rejects' (Jeffrey, 1956: 245). Once the scientist has assigned probabilities to hypotheses and communicated this information to 'rational agents' in society, these agents then assign utilities to the possible outcomes of accepting/rejecting the hypotheses and determine on the basis of a decision-theoretic calculation whether the hypothesis in question should be acted upon.

The primary argument that Jeffrey provides for the view that the scientist *qua* scientist does *not* accept or reject hypotheses is that the acceptance or rejection of a hypothesis *per se*, independent of a given practical context, is incoherent. In support of this, Jeffrey considers the following two hypotheses: (a) an entire lot of vaccine is safe, and (b) an entire lot of roller skate ball bearings is safe. In certain contexts, he argues, it is legitimate to demand a higher standard for the 'acceptance' of (a) than for (b) – but only when certain practical contexts are assumed, such as that the vaccine will be given to children. If we assume a different practical context, such as that the vaccine will be given to monkeys, the standards for 'acceptance' of the two hypotheses might be the same. Jeffrey summarizes his position by quoting approvingly the following passage from Bruno DeFinetti:

> I do not deem the usual expression 'to accept hypothesis H_r', to be proper. The decision does not really consist of this 'acceptance' but in *the choice of a definite action A_r*. The connection between the action A_r and the hypothesis H_r may be very strong, say 'the action A_r is that which we would choose if we knew that H_r was the true hypothesis'. Nevertheless, this connection cannot turn into an identification. (Quoted in Jeffrey, 1956: 242)

Thus, when one 'accepts' a hypothesis H, one is really choosing to act on the basis of H, and to do this requires that we specify a practical context in which the action is to occur. Because virtually all hypotheses could be acted upon in a multiplicity of different ways – e.g. the hypothesis that a given vaccine is safe could be used as a basis for vaccinating either children or monkeys – the notion of 'accepting' a hypothesis H *per se* is incoherent.

For our purposes, the primary implications of this view are (1) that one can distinguish clearly between the realms of theory and practice, or belief and action, and (2) that ethical considerations – or, in Jeffrey's terminology, considerations of utility – can be confined to the realm of practice. The scientist, Jeffrey argues, can and should remain in the realm of value-neutral theory and leave the ethical questions to 'the rational agents in the society which he represents'.

As stated earlier, Jeffrey's objection to Rudner's argument is still a standard objection to the view that ethical considerations play an inevitable role in the evaluation of research. For example, in response to Heather Douglas's argument that 'non-epistemic values are a required part of the internal aspects of scientific reasoning' (Douglas, 2000: 559), Sandra Mitchell maintains that this argument involves a 'conflation of the domains of belief and action [that] confuses rather than clarifies the appropriate role of values in scientific practice' (2004: 250). Moral and political values, according to Mitchell, play a legitimate role in the *practical*, not the *epistemic*, evaluation of research.

In the remainder of the chapter, we will argue that in the area of climate modeling, the Jeffreyan strategy does not succeed. In recent work, climate scientists, working alongside statisticians, have begun to attempt to do something very similar to what Jeffrey recommends; they attempt, that is, to estimate the uncertainties of various predictions made by climate models, and they attempt to do this in a manner that is free from moral, social, or any other kind of 'non-epistemic' value. They then hand these predictions and uncertainties over to policy-makers, legislators, and other representatives of the public, who are charged with determining how best to act.[7] In our view, however, this clean separation of the realms of value-neutral theory and value-laden practice is not realistically attainable. We will argue that one class of 'non-epistemic' values in particular – those that are reflected in decisions that certain types of prediction tasks are more important than others – influence the probabilities we assign to various possible climate outcomes. In order to argue for this, we need to establish a few preliminaries regarding the way in which uncertainties in climate modeling are estimated.

3. Climate modeling and uncertainty

Conceptually, we can distinguish three sources of uncertainty regarding the predictions of complex climate models. Firstly, there is uncertainty about the basic structure that our climate models ought to have. While the construction of climate models is guided by basic science, these models incorporate a barrage of auxiliary assumptions, approximations, and parameterizations, all of which contribute to a degree of uncertainty about the predictions of these models. We will call this type of uncertainty *structural model uncertainty*. Secondly, complex models involve large sets of parameters, or aspects of the model that have to be quantified before the model can be used to run a simulation of a climate system. We are often highly uncertain about what the best value for many of these parameters is, and hence, even if we had at our disposal a model with ideal (or perfect) structure, we would still be uncertain about the behavior of the real system we are modeling, because the same model structure will make different predictions for different values of the parameters. We will call uncertainty from this source *parameter uncertainty*. Finally, in evaluating a particular climate model, including both its structure and parameters, we compare the model's output to real data. Climate modelers, for example, often compare the outputs of their models to records of past climate. These records can come from actual meteorological observations or from proxy data – inferences about past climate drawn from such sources as tree rings and ice core samples. Both of these sources of data, however, are prone to error, and so we are uncertain about the precise nature of the past climate. This, in turn, has consequences for our knowledge of the future climate. We will call this source of uncertainty *data uncertainty*.

While data uncertainty is a significant source of uncertainty in climate modeling, we will not discuss this source of uncertainty here. For the purposes of this discussion, we will make the crude assumption that the data against which climate models are evaluated are known with certainty. We will be interested in arguing that values play an inevitable role in the estimation of uncertainties from the two other sources. How, then, do we estimate structural model uncertainty and parameter uncertainty? In both cases, there are, broadly speaking, two methods available to statisticians interested in quantifying these uncertainties. The first method makes use of *observable frequencies*, and the second *expert judgement*.

To understand the idea of using observable frequencies, consider the example of a simulation model with one parameter and several variables.[8] If one has a data set against which to benchmark the model, one could assign a weighted score to each value of the parameter based on how well it retrodicts values of the variables in the available data set. Based on this score, one could then assign a probability to each value of the parameter. Crudely speaking, what we are doing in an example like this is *observing* the *frequency* with which each value of the parameter is successful in replicating known data – how many of the variables does it get right? with how much accuracy? over what portion of the time history of the data set? – and then assigning this observed-frequency value to the probability of the parameter taking this value.

Of course, only in specific circumstances are frequencies probabilities. Absent other knowledge, it would naïve to think the observed frequencies in an example like the above are the actual probabilities of the values of those parameters. For one thing, we are interested in the best value of the parameter for predicting the behavior of the system for *all* times, not just the times for which we have sample data. For another, carrying out a procedure like the one above requires us to weight the relative importance of the various variables. Hence, while these kinds of frequencies can be useful guides in assigning probabilities to the values of a parameter, they are far from perfect, and some might view it to be more sensible to adopt a broadly subjectivist approach and to think that the best guide to these probabilities is the subjective degree of belief held by the best experts. Expert judgement, however, is not perfect either; in fact, expert judgement surely arises, *inter alia*, through the process of observing the degree to which model output matches available sample data. In practice, therefore, statisticians typically use some combination of expert judgement and observable frequencies to arrive at probabilities.

Thus, if we are interested in understanding where estimates of the degree of uncertainty about future climate come from, and, in particular, if we want to know to what degree these estimates are free from, or influenced by, various values, then we need to understand at least four things. We need to understand how observable frequencies and expert judgement are each used to estimate the degree of both structural model uncertainty and parameter uncertainty.

4. Structural model uncertainty

Let us begin with the estimation of structural model uncertainty. Given that one common method of estimating the degree of structural uncertainties in the predictions of climate models is by examining the degree of variation in the predictions of the existing set of climate models,[9] the range of models that happen to be available on the market will clearly influence our estimates of the degree of uncertainty contained in these models. In this section, we will argue that climate modelers have emphasized certain prediction and retrodiction tasks over others (e.g. predictions of global mean surface temperature change over predictions of global mean precipitation change), and that these past decisions have affected the performance of these models. We will then argue that this in turn invariably affects the current estimations of uncertainties. The decision to emphasize one prediction task over another is a paradigm example of a decision that reflects 'non-epistemic' values; that is, the decision is made not because emphasizing one prediction task over another has significant epistemic benefits, but rather because one set of prediction tasks is thought to be more important, in terms of its social, political, or economic consequences, than another. These value-laden decisions, we will argue, invariably affect the estimation of uncertainties of climate models.

As a preliminary to this argument, it is worth noting that the size of the dispersions of predictions of different climate models depends significantly upon the choice of a prediction task. Consider, for example, the graphs represented in Figure 9.1, which display predictions of temperature change and precipitation change made by different climate models.[10] For our purposes, what is important about these graphs is that the dispersion of predictions of mean precipitation change is significantly larger than the dispersion of predictions of global mean surface temperature change. The reason for this, we hypothesize, is that predictions of global mean surface temperature change have been more highly valued than other predictions; the climate modeling community has focused its energies upon refining and tweaking models in order to predict temperature change accurately, rather than precipitation change, sea level change, or any of a variety of other prediction tasks.

One might, however, question this hypothesis; after all, temperature and pressure are highly coupled variables in all of our models. Improvement in one should only be able to come in tandem with improvement in the other. Is there further evidence that the degree of structural model uncertainty that we get from 'observable frequencies' (from looking at the range of available models on the market) is affected by past values regarding the importance of various prediction tasks?

To answer this question, it is helpful to introduce what one might call the 'problem of attribution'. Since at least the late 1980s, the climate modeling community has been attempting to attribute the specific successes and failures of different climate models to specific components, or modules, of those models, for the purpose of developing models that predict all relevant quantities equally well. The thought was that once one understood the sources of disagreements of different climate models, one could take steps toward eliminating those disagreements. In 1989, a major program was founded at the Lawrence Livermore National Laboratory called the Program for Climate Model Diagnosis and Intercomparison (PCMDI), which had this as its stated goal. A number of projects were undertaken under the auspices of the PCMDI, including the Atmospheric Intercomparison Project, which began in 1990, and the Coupled Model Intercomparison Project, which began in 1995.[11]

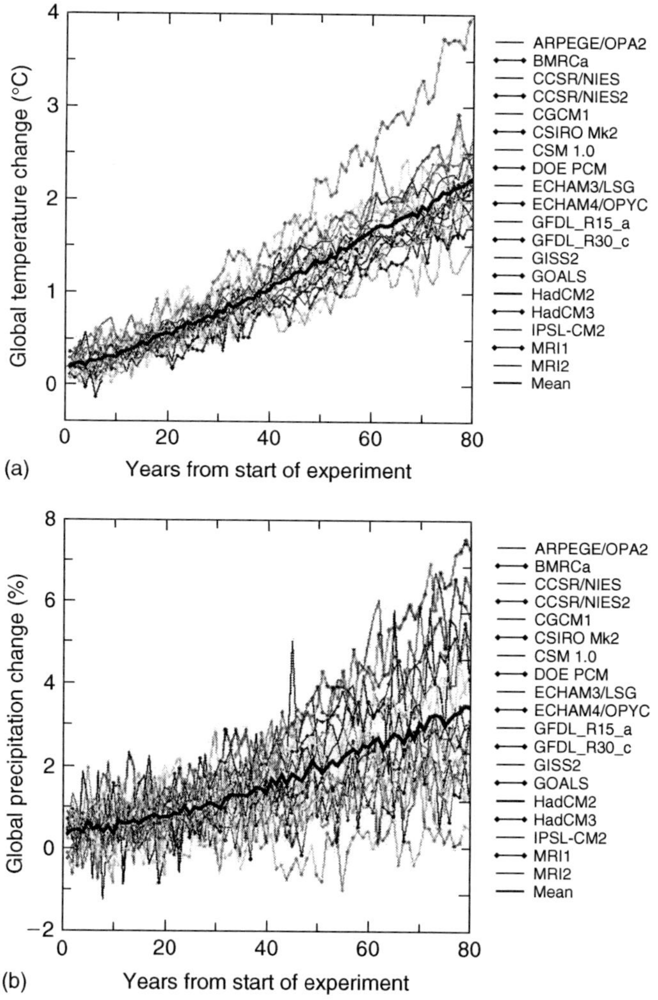

Figure 9.1 Graphs of global temperature change (°C) versus time (years) and global precipitation change (%) versus time (years)

Despite the optimism that surrounded the founding of the PCMDI, however, the project of attribution has been unsuccessful; not only this, there are strong reasons for believing that it will continue to be unsuccessful. Because the failure of the project of attribution is highly relevant to our argument that 'non-epistemic' values play an ineliminable role in the estimation of uncertainties, it will be necessary to discuss in some detail the reasons behind this failure.[12]

Kluges, generative entrenchment, and the historical character of model performance

The most sophisticated current climate models, Atmosphere-Ocean General Circulation Models (AOGCMs), are highly complex computer models that are constructed on the basis of both principled science – including fundamental partial differential equations from mechanics and thermodynamics – and trial and error approximations and parameterizations, and everything in between. An important feature of these models is their modular structure; each model is made up of components, or modules, and each module represents a given sub-system of the earth's climate, such as the circulation of the atmosphere, ice formation, ocean dynamics, cloud formation, the effects of vegetation, and the dynamics of aerosols. Today's climate models have their roots in the General Circulation Models of the 1950s, which described the atmosphere; over time, the complexity of these models has increased via the addition of more and more modules. The addition of these modules has allowed for better predictions of a growing range of phenomena; as already indicated, however, the predictions made by different climate models differ from one another, in some cases significantly.

An important reason for this variation in predictions is that the modules that make up a climate model are not mutually independent, but rather interact with one other, such that the overall performance of a model depends not simply upon the representational adequacy of each individual module, but also upon the way in which the modules are coupled together. Thus, if module X is good at predicting phenomenon Y, and if module W is good at predicting phenomenon Z, it will not, at least in general, be the case that the model consisting of X and W will be good at predicting both Y and Z. This overall model might be good at predicting one or the other of these phenomena, or neither, depending upon the way in which the modules interact with one another.

Moreover, not only does the overall performance of a given climate model depend upon the interaction of its modules; it is also the case that the specific ways in which modules interact with one another will depend upon the overall model of which they are 'parts'. The reason for this is that the process of coupling modules together relies to a large extent upon fitting the module to the overall model – not only to its principled structure but also its parameterizations.[13] This, in turn relies on trial and error, piecemeal mutual adjustments of parameters and parameterization schemes – adjustments that need to be undertaken in different ways, depending upon the details of the model and modules in question. In order to illuminate the nature of these adjustments, a number of commentators have employed the notion of a 'kluge', a colloquial term first employed by computer programmers to describe sections of code that were functional but unprincipled, inelegant, and ill-understood. According to Andy Clark, a kluge is 'an inelegant, "botched together" piece of program; something functional but somehow messy and unsatisfying' (Clark, 1987: 278). The term describes aptly the process of coupling two modules together; while parts of the process might be principled, there are other parts that rely to a large extent upon simply botching things together.

In our view, the rather messy character of module coupling has important implications for the development of climate models; it suggests that model development is an historical process that depends in important respects upon the environment in which that model develops. To see this, it is helpful to draw upon William Wimsatt's notion of 'generative entrenchment' (Wimsatt, 2007). According to Wimsatt, 'a deeply generatively entrenched feature of a structure is one that has many other things depending on it because it has played a role in generating them' (2007: 133). Wimsatt employs this notion in order to explain the relationship between biological development and evolution; characteristics that are adaptive for one organism will not, in general, be adaptive for another, because the different organisms will, in general, have different features that are generatively entrenched. Unlike Dumbo, real elephants will never be able to fly, because they have particular features – e.g. bulkiness – that have developed in specific circumstances as a result of specific environmental pressures and that make adaptations such as wings impossible. Bulkiness is a generatively entrenched feature of elephants. Analogously, there are some climate models that will never be successful at predicting a particular phenomenon via a particular module, because the model has developed in specific circumstances as a result of specific 'environmental' pressures – e.g. pressures to emphasize certain predictive tasks over others – that make a given module impossible. A particular method of kludging the model might be as generatively entrenched as bulkiness is to an elephant. Just as tusks won't help a bird and wings won't help an elephant, a module that helps model *A* might not help model *B*.

Consider the following idealized example. Climate Model *A* is a simple General Circulation Model that describes the dynamics of the atmosphere. By building upon *A*, two further models, *B* and *C*, develop, both of which describe the following sub-systems of the earth's climate: atmosphere dynamics, ocean dynamics, ice formation, and the effects of vegetation. Model *B* develops from *A* via the following path: first, a module for ocean dynamics is coupled to Model *A*, followed by a module for ice formation, and finally a module for the effects of vegetation. Model *C* took a slightly different path; first, a module for ocean dynamics is coupled to *A*, followed by a module for the effects of vegetation, and finally a module for ice formation. Given the fact that *B* and *C* took different developmental paths, the attempt to couple additional modules to *B* and *C* – e.g. a module for aerosols – will, in general, need to proceed along different lines for each model; the piecemeal approximations to parameterization schemes that will allow a new module to be coupled to *B* will, in general, not work for *C*, and vice versa. This does not mean that it will be impossible to add an aerosol module to *C*; however, given the ways in which *B* and *C* have developed, it might be significantly more difficult to add this module to *C* than to *B*, and it almost certainly will need to be done via a very different coupling process. These difficulties, moreover, only increase with the complexity of the model in question. Thus, given models of the complexity that we have today, the historical development of these models – which, again, has been influenced significantly by the 'environmental' pressures placed upon them – places constraints upon the kinds of modules that can be coupled to them, which in turn affects the overall performance of the models.

The preceding discussion should make clear why the problem of attribution has been, and will in all likelihood continue to be, unsuccessful. One cannot attribute the predictive success or failure of a given model to a particular, localized component of that model because the components of models are strongly coupled to one another, and hence interact with one another in significant fashion. Moreover, the kinds of additions that can be made to a climate model – e.g. the ways in which the scope of a model can be expanded – will in general depend upon the way in which that model has developed over time, including the approximations and adjustments that have been made in order to couple new modules successfully. This, in turn, will depend upon the 'environmental' pressures to which the model is subject. All of this suggests that previous decisions to emphasize certain prediction and retrodiction tasks over others – e.g. predictions of global mean surface temperature change over predictions of global precipitation change – have a significant effect upon the ways in which these models can be expanded, and upon the ways that future expansions of these models will perform.

Values and the estimation of structural model uncertainty

What does all of this have to do with the debate over the role of values in scientific research? According to the preceding discussion, the overall performance of a model – its ability to predict and retrodict certain tasks well, and its inability to predict and retrodict other tasks well – depends to a significant extent upon the history of that model, which in turn depends upon that model's 'environment', which includes political decisions to emphasize certain predictive and retrodictive tasks over others. Because the estimation of uncertainties of our knowledge of climate change depends upon the performance of our best climate models, and because decisions of a 'non-epistemic' character have an ineliminable effect upon the performance of our best climate models, such 'non-epistemic' values invariably influence the estimation of uncertainties. Thus, Jeffrey's claim that we can assign probabilities to hypotheses in a value-neutral fashion – or, in the present case, that we can assign uncertainties to predictions in a value-neutral fashion – is false, at least in the area of climate modeling.

There are a number of ways that one might object to this argument. First of all, our argument depends upon the claim that the choice of prediction task is inevitably influenced by 'non-epistemic' factors; yet one might argue that the decision to emphasize temperature predictions over other predictions can be explained solely on theoretical, or 'purely epistemic', grounds. While there are a variety of ways in which one could develop this objection, perhaps the most plausible would proceed as follows. Of the candidate prediction tasks that one could choose to emphasize – including global mean surface temperature change, ocean heat content change, polar ice cap retreat, and so on – temperature is the most significant theoretically, because all of the other quantities can be derived from it. Without knowing how global mean surface temperature will change, one cannot explain or make sense of any other change in global climate; therefore, predictions of temperature change should be emphasized.

The objection thus formulated, however, has its problems. It is true that CO_2 and other gases are often called 'heat trapping', but it would be more accurate to call them 'energy trapping', and one of the most difficult problems within contemporary climate research is determining where this energy will go. Will it go towards heating the ocean? Melting sea ice? Or raising temperature? If it goes towards the latter, will it do this on the surface of the earth or in other parts of the atmosphere? To maintain that all of these other features of the planet's dynamics can be *derived* from mean global surface temperature is misguided, because it fails to take into account the massive and not entirely understood interdependencies that exist in the climate. All of these interdependencies together react in complex fashion to the trapping of additional energy, and it would be a mistake to assume that this additional energy will always manifest itself as higher surface temperatures. Increased surface temperature is only one of many symptoms of global climate change.

It seems to us that this is the most plausible way of developing this objection, and yet it still fails. One could, of course, attempt to formulate it in another fashion. We are skeptical, however, that such a route will be successful. Until this burden is met, therefore, we conclude that the decision to emphasize predictions of global mean surface temperature change is one that is ineliminably influenced by 'non-epistemic' considerations, such as the fact that this prediction task was seen as the most important, in terms of its social, political, and economic consequences.

A second objection that one might direct against our argument is that the way in which values operate in this case is philosophically uninteresting; if scientists decide, for whatever reason, to put more resources into the prediction of one task than another, it is hardly novel to maintain that the resulting models will be better at predicting the former task than the latter. This objection, however, fails for at least two reasons.[14]

First of all, the objection obtains its intuitive force via a false presupposition, namely that it is realistic to attempt to construct models that perform in a manner that is independent of the influence of value judgements. According to the previous discussion, the performance of current climate models depends *invariably* upon the historical development of those models, which in turn depends *invariably* upon decisions of a social or political character, including decisions that a given prediction and retrodiction task is more important, from a social, moral, or economic point of view, than another. If it were reasonable to expect the performance of climate models to be independent of such past judgements, then the objection would be a good one; according to our argument, however, this expectation is not at all reasonable.

Secondly, and perhaps more importantly, the objection misrepresents the effects of focusing upon temperature predictions at the expense of precipitation predictions. In particular, the objection suggests that the primary effect of focusing upon one kind of prediction task over another is a difference in the actual uncertainties associated with these two predictive tasks. Yet, it is not merely the case that a lack of focus upon precipitation predictions will result in models that perform poorly with respect to this prediction task – and thus to higher actual uncertainties about future precipitation; it is also the case that a focus on mean global surface temperature predictions at the expense of global precipitation predictions will likely result in a range of models that *overstates* our estimates of the uncertainty of precipitation predictions and *understates* the uncertainty of temperature predictions. This choice of prediction tasks, in other words, affects not only the actual uncertainties of the models in question, but also the *estimations* of these uncertainties.

The reason for this pertains to the problems associated with estimating structural model uncertainty via observable frequencies – or, more specifically, via observing the range of predictions generated by our existing arsenal of models. If there are some particularly bad models on the market, we will overestimate the degree of uncertainty that we ought to have about our best models. On the other hand, if there is some degree of co-evolution in our models – that is, if the makers of our models deliberately ensure that the predictions of their model do not deviate too much from the herd (a natural thing to do if you do not want it to seem like the predictions of your model are unrealistic) – then the range of models on the market will underestimate the degree of uncertainty that we ought to attribute to them. As Myles Allen, who refers to this method as using 'ensembles of opportunity', notes: 'If modelling groups, either consciously or by "natural selection", are tuning their flagship models to fit the same observations, spread of predictions becomes meaningless: eventually they will all converge to a delta-function' (Allen, 2008). Thus, by misrepresenting the effects of focusing upon temperature predictions at the expense of precipitation predictions, the objection obscures the significant impact that values have in this situation.

Structural model uncertainty and expert judgement

Of the four modalities that we discuss in this chapter (expert judgement versus observable frequencies and structural model uncertainty versus parameter uncertainty), we have thus far restricted our attention to the estimation of structural model uncertainty via observable frequencies. The aforementioned problems with this method, however, lead to a further challenge to our claim that 'non-epistemic' considerations play an ineliminable role in the estimation of structural model uncertainty – namely that one could perhaps make such estimations via expert judgement, rather than observable frequencies, and do so in a manner that is free from 'non-epistemic' values. Are there reasons for believing that values play an ineliminable role in the case of expert judgement as well?

While we believe that there are such reasons, the question is nevertheless rather difficult to answer, primarily because, of the four modalities under consideration, the use of expert judgement to estimate structural model uncertainty is the one with the least developed methodology in the literature.[15] The source of this lacuna is clear; when statisticians solicit the opinions of experts regarding the degree of structural uncertainty in a model, the statisticians are often confronted with the problem that the experts are either unwilling or unable to assign subjective degrees of belief to the structure of a model. In particular, modelers are extremely reluctant to assign probabilities to the predictions of models that are deterministic, given a particular value in parameter space.[16] When statisticians, trying to implement this modality, make it clear that what they are after is the modelers' subjective degree of belief, the modelers are apt to balk. Thus, it is not at all clear that the relevant experts actually possess the subjective degrees of belief, and methods need to be created that will reliably elicit these judgements.

The issue that is of interest to us, of course, is the extent to which experts' judgements about their subjective degrees of belief, assuming that they are willing to make these judgements, are influenced by past decisions about prediction tasks. Given, again, that a well-established methodology for making these judgements does not yet exist, the issue is a difficult one to resolve. We can, however, make the following preliminary argument. In order for expert judgement to overcome the effects that past prediction task priorities have had on ensembles of existing models, the experts in question will have to understand exactly what these effects have been. Yet, the failure of the project of attribution, discussed above, leads us to be skeptical that one can know what the precise effects of past prioritizations upon existing models actually have been. This, in turn, makes us skeptical that expert judgement can be free from 'non-epistemic' considerations, such as the choice of one prediction task over another.

Where does this leave us? We have argued that estimations of structural model uncertainty via observable frequencies cannot be done in a value-free way, and we have called into question the claim that this can be done in the case of expert judgement as well. In order to establish this latter point more firmly, more research on the methodology of this modality needs to be done. In the meantime, we have to be content with the fact that the predominant method in actual use in the scientific community is to look at observable frequencies; this method, as we have argued, is not value free.

Suppose, however, that one is unconvinced by the arguments just provided. Even if one continues to maintain that the estimation of structural model uncertainty can be done independently of 'non-epistemic' considerations – indeed, even if one assumes that there is no uncertainty whatsoever about the structure of a model – there is still significant uncertainty about the values of parameters and parameterization schemes. We believe that there is an even stronger case to be made for the claim that 'non-epistemic' considerations affect the estimation of parameter uncertainty. It is to this argument that we will now turn.

5. Parameter uncertainty

To keep the issue of parameter uncertainty conceptually distinct from that of structural model uncertainty, let us suppose that we know that we have a structurally perfect model with n parameters and that we are uncertain about what the best value is for each of these parameters. We can think of the set of n parameters as forming an n-dimensional parameter space. If we had a probability density function (PDF) over that n-dimensional space, we could easily assign probabilities to the predictions of the model. In principle, we would use something like the following Monte Carlo method: we could sample from the space of parameters in accordance with the given PDF, and calculate the output of the model for each of the sampled points in the space.[17] The resulting set of outputs would have means and variances for each of the variables that would correspond to the uncertainties we would assign to the corresponding predictions. If, for example, the mean prediction of the probability weighted sample for temperature is x, with standard deviation of s, then we could say that it is 95 percent likely that the modeled system's temperature will be $x \pm 2s$.

The question, then, is where a PDF over the space of possible parameter values can come from. The answer, as noted, is from observed frequencies and expert judgement. We've already seen, in some detail, how the former works. Each value in parameter space can be benchmarked; it can be scored for its ability, when used as input for a given model structure, to reproduce existing data, and probability densities can be assigned in proportion to that score. It is important to remember, however, that climate models are highly multivariate in their output. Scoring the performance of each point in parameter space, therefore, requires us to weight the relative importance of making good predictions of each variable. Such a weighting involves a value decision that involves both 'epistemic' and 'non-epistemic' considerations.[18] More importantly, expert judgement might very well affirm that a point in parameter space that is very successful at reproducing known, available data will be poor at predicting unknown, unavailable data. In practice, therefore, a more common procedure is to use a benchmarking procedure to rule out certain regions of parameter space – if a point in parameter space does a very poor job of reproducing known data then it can be thrown out – and then use expert judgement to assign a PDF to the remaining subset of the space.

But what are experts judging when they assign such a PDF? Here, we think it is important to keep in mind an important difference between two kinds of parameters that climate models can take as inputs. So far, we have simply defined a parameter of a model as any aspect of the model that has to be quantified before the model simulator can be run. But the aspect of the model being quantified might or might not correspond to some actual quantifiable or measurable property of the physical system being modeled. While some of the parameters of our models are of a kind with physical parameters like g, the acceleration due to gravity near the surface of the earth ($9.8\,m/s^2$), others of them, which result from so-called 'parameterizations' in the model, are rather different.

Parameterizations are elements of a simulation model that are designed to capture effects that slip between the cracks of a model's discretization grid or are otherwise lost to an approximation of the model.

> Sub-grid processes are represented by *parameterisation schemes* describing their aggregated effect over a larger scale. These schemes are often referred to as 'model physics' but are really based on physics-inspired statistical models describing the mean quantity in the grid box, given relevant input parameters. (Benestad, 2007)

Parameterization schemes are extremely common in global climate models, especially given that reductions in discretizations and approximations are bought at the price of complexity of computer models, increase of simulation time, and so on (ibid.).

In our view, there is an important difference between these two different kinds of parameters. With respect to the first kind of parameter, it makes sense to talk about the correct value of the parameter. For example, near the surface of the earth, the correct value for g is approximately $9.8\,m/s^2$. But the value of a parameter associated with a parameterization scheme does not have a single correct value. At best, it has a best value for a particular prediction task – a value which, if used as model input, will enable the model to make the best possible predictions for that particular task.

Goldstein and Rougier highlight the important difference between these two kinds of parameters when it comes to estimating uncertainties.

In their discussion of eliciting expert judgement about appropriate PDFs to place on parameter space, they write:

> In practice, modellers often seem to take two somewhat contradictory positions about the status of the simulator's best input, on the one hand arguing that it is a hypothetical construct and on the other hand using knowledge and intuition derived from the physical system to set plausible intervals within which such a value should lie. (Goldstein and Rougier, 2006: 5)

Of course, when it comes to parameterization schemes, there is no straightforward relation between the values of the inputs to the simulator and any measurable corresponding physical values for the system. If expert judgement is to be relied upon in assigning probabilities to parameter values, therefore, it has to come not from *knowledge and intuition of the physical system alone*, but from knowledge and intuition of the behavior of the model vis-à-vis the system. In that regard, one more passage from the Goldstein and Rougier piece is particularly striking: 'In particular, there was strong disagreement [among statistically numerate system experts] with [the following:] that there exists an input x* such that, were it to be known, only a single evaluation of the simulator would be necessary' to learn everything that the model has to tell us about the system (ibid.). What the statistically numerate system experts are suggesting is that the best value from parameter space for one particular prediction task is not necessarily the same as the one for another prediction task. Indeed, it may even be that for some prediction task, the maximum amount of knowledge that can be extracted from a particular model structure might come from an ensemble of runs from multiple parameter values. Thus, the idea that there is a single 'best value' for a parameter in a model is, in many cases, not correct. The best value for prediction task A might not be the same as for prediction task B. And even for a single prediction task, it might be an ensemble of different parameter values that yields the most valuable knowledge.

What is important to note here is that, as emphasized in the previous section, climate models can be applied to a variety of prediction tasks. While climate modelers have traditionally emphasized predictions of global mean surface temperature change, they have more recently begun to turn their attention to other prediction tasks, such as the probability of so-called 'abrupt' climate changes, including the collapse of thermohaline circulation, the drying of the Amazon, and the disappearance of Arctic summer sea ice. What we do not know for a particular climate model with all of its parameterizations, is whether the best parameter value for predicting mean surface temperature is also the best parameter value for predicting, say, the extent of thermohaline circulation. But if the claim made by Rougier and Goldstein's statistically numerate system experts is right (and we think it is), and they are not always the same, then the following suspicion arises.

Suppose we are confronted by a novel prediction task, such as estimating the probability of thermohaline collapse. And suppose that, for a particular model structure, we want to estimate the degree of parameter uncertainty about such predictions. To do this, as we have discussed, would require us to use expert judgement to assign a PDF to the parameter space for that model. If what experts are judging is where, in parameter space, the 'best' value of parameters can be found, and if the best value for predicting mean surface temperature is not necessarily the best value for predicting thermohaline circulation, then the suspicion is that expert judgement about parameter space PDFs will be influenced, to some degree or other, by the intended purposes for which these experts have been constructing and testing their models during the period in which they have acquired their expertise. Again, when it comes to judgement about parameterizations, the relevant expertise is not about the system itself, but about the behavior of the model vis-à-vis the system.

Values and the estimation of parameter uncertainty

At this point, we are in a position to state the argument for the ineliminability of 'non-epistemic' considerations in the estimation of parameter uncertainty. Suppose that we have a particular model structure, about which there is no uncertainty, and suppose furthermore that we want to employ this model in order to estimate the probability of thermohaline collapse given a doubling of atmospheric carbon dioxide. To estimate this, we have to elicit expert opinion regarding an appropriate PDF to assign to the space of parameters. But the best value(s) of parameters for predicting thermohaline circulation is not necessarily the same as the best value(s) for other prediction tasks, such as predicting mean surface temperature. If the experts whose opinion we elicit have predominantly acquired experience with their models in predicting mean surface temperature, then this will affect their judgement about where the best value for parameters can be found.

But the particular set of prediction tasks that have played a role in shaping our experts' judgements have been the product of a set of choices – for example, the choice to focus on predicting mean surface temperature rather than mean global precipitation. And these choices, in turn, reflect a set of values – namely the set of social, economic, or other considerations that have historically led us to believe that predicting temperature is more important than predicting precipitation. These values, in other words, are ones that are traditionally regarded as being 'non-epistemic' in character. Thus, 'non-epistemic' values play an ineliminable role in the estimation of parameter uncertainty.

6. Conclusion

Jeffrey argues that the task of the scientist can and should be limited to the assignment of probabilities to hypotheses, and that this task can be carried out in a manner that is free from practical, 'non-epistemic' considerations. Once the task of assigning probabilities to hypotheses has been completed, the scientist can then hand these hypotheses and probabilities over to policy-makers, who are responsible for deciding upon a course of action. On this picture, there is a clean separation between the realms of theory and practice and a clear line that divides the space where values play a legitimate role (the realm of practice) and where they do not (the realm of theory). As noted earlier, Jeffrey's line of reasoning is still very commonly followed today.

If our argument is sound, however, Jeffrey's line of reasoning fails, at least in one very important area of contemporary scientific research – namely, climate modeling. Scientists cannot assign probabilities to hypotheses about climate change – or, more specifically, estimate the uncertainties of climate predictions – in a manner that is free from 'non-epistemic' considerations, because 'non-epistemic' considerations invariably influence the choices of prediction tasks, and the choices of prediction tasks invariably influence the estimation of both structural model uncertainty and parameter uncertainty.

Again, we do not believe that this result in any way implies that the consensus that has been formed regarding the causal connection between fossil fuel emissions and global climate change is problematic. We are not skeptics of climate change. We do, however, believe that this conclusion suggests that more attention should be paid to the spaces within climate modeling where values play a role, to the kinds of values or 'non-epistemic' considerations that play a role, and to the effects that these values have upon the overall performance of our models.

Notes

*We have accrued many debts in writing this chapter. Thanks to Heather Douglas and Elizabeth Lloyd for their comments on an earlier draft. Thanks to James

McAlister, Michael Weisberg, Ron Giere, and others for their comments on presented versions of this paper in Dubrovnik and Mexico City. We began thinking about this topic as Fellows of the research group, Science in the Context of Application, at the Center for Interdisciplinary Research (ZiF) at Bielefeld University. Thanks to Martin Carrier and Alfred Nordmann for organizing this group, and thanks to Johannes Lenhard, Torsten Wilholt, Cornelis Menke, and Justus Lentsch for listening to some early ideas. Special thanks to Johannes for allowing us to borrow liberally from his and Eric's collaboration. Eric would like to thank the IAS, Durham University, for giving him time to work on this chapter, and for exposing him to the expertise of several statisticians who work on climate models, especially Michael Goldstein, Peter Challenor, and John Haslett.

1. Defenders of this view include Giere (2003), Jeffrey (1956), Kitcher (2001), Koertge (2003), McMullin (1983), and Mitchell (2004).
2. It has not always been standard to allow a role for values of any kind in the appraisal of theories. One of the primary aims of the neo-positivist movement in post-Second World War America was to explicate fundamental methodological concepts such as confirmation and explanation in purely formal terms (e.g. Hempel, 1945; Hempel and Oppenheim, 1948). Had this project succeeded, there would be no need for any values – including epistemic values – in the appraisal of theories. The project, however, did not succeed, and since the demise of neo-positivism, it has been standard to argue that values of some sort play an inevitable role in theory appraisal. For classic expressions of this view, see Kuhn (1969, 1977) and McMullin (1983).
3. See, for example, Douglas (2000), Howard (2006), Longino (1990, 2002), Kourany (2003a, 2003b), Solomon (2001), and Wilholt (forthcoming).
4. For example, Ron Giere employs a version of this objection in his criticism of Janet Kourany's proposal for a socially responsible philosophy of science (Giere, 2003; Kourany, 2003a, 2003b), and Sandra Mitchell employs it against the arguments of Heather Douglas (Douglas, 2000, 2004a, Mitchell, 2004).
5. Throughout this chapter, we will employ the traditional terminology of 'epistemic' and 'non-epistemic' values. We recognize, however, that this terminology is in many respects problematic, as has been argued in Longino (1996), Rooney (1992), and a number of the works cited in note 3. As is apparent from our previous discussion, the terminology arose in the context of arguments to the effect that social values, moral values, and other broadly practical considerations can and should be excluded from the epistemic evaluation of research – hence the label 'non-epistemic'. If it turns out, however, that such values do play a legitimate role in the epistemic evaluation of research, then it makes little sense to label them 'non-epistemic' values. Thus, our essay – in addition to the literature cited in note 3 – provides strong reasons for abandoning the terminology of 'epistemic' and 'non-epistemic' values. However, given the entrenched character of this terminology, we have decided in the interests of convenience to use it here.
6. See Douglas (2004a), Fine (1998), and Longino (1990) for discussions of the compatibility of values and objectivity in science.
7. For a clear example of this, see the so-called 'Stern Review' (Stern, 2007).
8. A parameter for a model is an input that is fixed for all time, while a variable takes a value that varies with time. A variable for a model is thus both an

input for the model (the value the variable takes at some initial time) and an output (the value the variable takes at all subsequent times). A parameter is simply an input.

9. See, for example, IPCC (2007) or, for a more skeptical position regarding the feasibility of this endeavor, Smith (2002). The typical method is simply to take the average and standard deviation of the existing predictions and generate uncertainties from these values.
10. Taken from IPCC (2001: 537).
11. For further details, see the PCMDI website: http://www-pcmdi.llnl.gov/about/index.php (accessed May 13, 2008).
12. The following discussion draws heavily upon Winsberg and Lenhard (forthcoming).
13. Parameterizations, roughly speaking, are correction factors, or attempts to correct for the fact that simulation models cannot capture effects that occur at scales below which the models are discretized. We discuss parameterizations in much greater detail in section 5.
14. Thanks to James McAllister for encouraging us to think further about this objection.
15. For a discussion of these methods, see Goldstein and Rougier (2006). For criticism of these methods, see Allen (2008).
16. Modelers, in fact, often think statisticians will do that for them. Statisticians, on the other hand, clearly (and correctly) view this as a scientific problem whose answer can only be provided by those with the best scientific expertise.
17. We say 'in principle' because in practice the situation is more complicated. In practice, it is too time consuming to calculate the outputs of a global climate model for sufficiently many sampled points. Statisticians, therefore, use so-called 'emulators' of the models, in effect simulations of the simulations, to calculate these outputs. While this introduces a large layer of technical difficulties to the project, it has no bearing, that we are aware of, on what we discuss here.
18. Here, we think, is an interesting example of where the purported distinction between epistemic values and non-epistemic values breaks down. Whether or not the value underlying such a weighting would count as epistemic or non-epistemic would depend on whether the researchers could argue that their choice of weighting was made on the basis of which weighting was the 'best guide to truth'. But which truths? And how would such an argument be resolved? The distinction here becomes a bit confused.

References

Allen, M. (2008) 'What Can be Said about Future Climate?' http://www.climateprediction.net/science/pubs/allen_Harvard2008.ppt (accessed July 3, 2008).

Benestad, R. (2007) 'Why Global Climate Models Do Not Give a Realistic Description of the Local Climate', *RealClimate*, http://www.realclimate.org/index.php/archives/2007/05/climate-models- local-climate (accessed May 29, 2007).

Churchman, C. West (1948) *Theory of Experimental Inference* (New York: Macmillan).

Churchman, C. West (1956) 'Science and Decision Making', *Philosophy of Science*, 22: 247–249.

Clark, A. (1987) 'The Kluge in the Machine', *Mind and Language*, 2: 277–300.
Dewey, J. (1929) *The Quest for Certainty* (New York: Minton, Balch & Company). Reprinted in J. A. Boydston (ed.), *John Dewey: the Later Works, 1925–1953*, vol. IV (Carbondale and Edwardsville: Southern Illinois Press, 1988).
Douglas, H. (2000) 'Inductive Risk and Values in Science', *Philosophy of Science*, 67: 559–579.
Douglas, H. (2004a) 'Border Skirmishes between Science and Policy: Autonomy, Responsibility, and Values', in P. Machamer and G. Wolters (eds), *Science, Values, and Objectivity* (Pittsburgh: University of Pittsburgh Press), 220–244.
Douglas, H. (2004b) 'The Irreducible Complexity of Objectivity', *Synthese*, 138: 453–473.
Fine, A. (1998) 'The Viewpoint of No-One in Particular', *Proceedings and Addresses of the American Philosophical Association*, 72: 9–20.
Frank, P. G. (1954) 'The Variety of Reasons for the Acceptance of Scientific Theories', in P. G. Frank (ed.), *The Validation of Scientific Theories* (Boston: Beacon Press), 3–17.
Giere, R. N. (2003) 'A New Program for Philosophy of Science?', *Philosophy of Science*, 70: 15–21.
Goldstein, M. and J. Rougier (2006) 'Reified Bayesian Modelling and Inference for Physical Systems', reprint for *Journal of Statistical Planning and Inference*.
Hempel, C. G. (1945) 'Studies in the Logic of Confirmation', *Mind*, 54: 1–26, 97–121. Reprinted in Hempel (1965), 3–46, with a 1964 Postscript added.
Hempel, C. G. (1965) *Aspects of Scientific Explanation and Other Essays in the Philosophy of Science* (New York: The Free Press).
Hempel, C. G. and P. Oppenheim (1948) 'Studies in the Logic of Explanation', *Philosophy of Science*, 15: 135–175. Reprinted in Hempel (1965), 245–290, with a 1964 Postscript added.
Howard, D. A. (2006) 'Lost Wanderers in the Forest of Knowledge: Some Thoughts on the Discovery–Justification Distinction', in J. Schickore and F. Steinle (eds), *Revisiting Discovery and Justification: Historical and Philosophical Perspectives on the Context Distinction* (New York: Springer).
IPCC (2001) *Climate Change 2001: the Scientific Basis. Contribution of Working Group I to the Third Assessment Report of the Intergovernmental Panel on Climate Change* (Cambridge: Cambridge University Press).
IPCC (2007) *Climate Change 2007: the Physical Science Basis. Contribution of Working Group I to the Fourth Assessment Report of the Intergovernmental Panel on Climate Change* (Cambridge: Cambridge University Press).
Jeffrey, R. C. (1956) 'Valuation and Acceptance of Scientific Hypotheses', *Philosophy of Science*, 22: 237–246.
Kitcher, P. (1993) *The Advancement of Science: Science without Legend, Objectivity without Illusions* (New York: Oxford University Press).
Kitcher, P. (2001) *Science, Truth, and Democracy* (New York: Oxford University Press).
Koertge, N. (2003) 'Screen Out Social Values from Core Epistemic Areas of Science', in C. Pinnick, R. Almeder, and N. Koertge (eds), *Scrutinizing Feminist Epistemology: an Examination of Gender in Science* (New Brunswick, NJ: Rutgers University Press).
Kourany, J. (2003a) 'A Philosophy of Science for the Twenty-First Century', *Philosophy of Science*, 70: 1–14.

Kourany, J. (2003b) 'Reply to Giere', *Philosophy of Science*, 70: 22–26.

Kuhn, T. (1962) *The Structure of Scientific Revolutions* (Chicago: University of Chicago Press).

Kuhn, T. (1969) 'Postscript – 1969', in *The Structure of Scientific Revolutions*, 2nd edn (Chicago: University of Chicago Press), 174–210.

Kuhn, T. (1977) 'Objectivity, Value Judgment, and Theory Choice', in *The Essential Tension: Selected Studies in Scientific Tradition and Change* (Chicago: University of Chicago Press), 320–339.

Longino, H. (1990) *Science as Social Knowledge: Values and Objectivity in Scientific Inquiry* (Princeton, NJ: Princeton University Press).

Longino, H. (1996) 'Cognitive and Non-Cognitive Values in Science: Rethinking the Dichotomy', in L. H. Nelson and J. Nelson (eds), *Feminism, Science, and the Philosophy of Science* (Dordrecht: Kluwer Academic Publishers), 39–58.

Longino, H. (2002) *The Fate of Knowledge* (Princeton, NJ: Princeton University Press).

Machamer, P. and G. Wolters (eds) (2004) *Science, Values, and Objectivity* (Pittsburgh: University of Pittsburgh Press).

McMullin, E. (1983) 'Values in Science', in P. D. Asquith and T. Nickles (eds), *Proceedings of the 1982 Biennial Meeting of the Philosophy of Science Association*, vol. I (East Lansing, MI: Philosophy of Science Association), 3–28.

Mitchell, S. D. (2004) 'The Prescribed and Proscribed Values in Science Policy', in P. Machamer and G. Wolters (eds), *Science, Values, and Objectivity* (Pittsburgh: University of Pittsburgh Press), 245–255.

Neurath, O. (1913). 'Die Verirrten des Cartesius und das Auxiliarmotiv. Zur Psychologie des Entschlusses', in *Jahrbuch der Philosophischen Gesellschaft an der Universität Wien* (Leipzig: Johann Ambrosius Barth). English translation: 'The Lost Wanderers of Descartes and the Auxiliary Motive (On the Psychology of Decision)', in Otto Neurath, *Philosophical Papers 1913–1946*, ed. and trans. R. S. Cohen and M. Neurath (Dordrecht: D. Reidel), 1–12.

Polanyi, M. (1962) 'The Republic of Science: Its Political and Economic Theory', *Minerva*, 1: 54–74.

Rooney, P. (1992) 'On Values in Science: Is the Epistemic/Non-Epistemic Distinction Useful?', in D. Hull, M. Forbes, and K. Okruhlick (eds), *Proceedings of the 1992 Biennial Meeting of the Philosophy of Science Association*, vol. II (East Lansing, MI: Philosophy of Science Association), 13–22.

Rudner, R. (1953) 'The Scientist *Qua* Scientist Makes Value Judgments', *Philosophy of Science*, 20: 1–6.

Smith, L. A. (2002) 'What Might We Learn from Climate Forecasts?', *Proceedings of the National Academy of Sciences USA*, 4(99): 2487–2492.

Solomon, M. (2001) *Social Empiricism* (Cambridge, MA: MIT Press).

Stern, N. (2007) *The Economics of Climate Change: the Stern Review* (Cambridge: Cambridge University Press).

Wilholt, T. (forthcoming) 'Bias and Values in Scientific Research', *Studies in History and Philosophy of Science*.

Wimsatt, W. (2007) *Re-engineering Philosophy for Limited Beings: Piecewise Approximations to Reality* (Cambridge, MA: Harvard University Press).

10

Feminist Standpoint Empiricism: Rethinking the Terrain in Feminist Philosophy of Science

Kristen Intemann

1. Introduction

Feminist philosophy of science involves two main projects. The first is identifying how gender (and other categories of oppression) can influence science (including what is studied and taken to be scientific knowledge). The second, more normative, project is to develop the implications this has for our epistemic concepts and practices. For example, feminist philosophers of science are concerned with how to understand objectivity (given the ways that gender can influence science), minimize biases, and promote objectivity.

In her 1986 *The Science Question in Feminism*, Sandra Harding identified the views that are still seen as constituting feminist philosophy of science today. In particular, she distinguished and endorsed 'standpoint feminism' over 'feminist empiricism', which she described as the view that instances of male bias are merely cases of 'bad science' that could be eliminated if scientists more rigorously adhered to empiricist methods and norms for scientific research (Harding, 1986, 1991: 111–120).

Since that time, however, increasingly sophisticated forms of feminist empiricism have emerged, which do not fit Harding's characterization (Anderson, 1995, 2004; Campbell, 1998; Hankinson Nelson, 1990; Longino, 1990). Also, many self-described standpoint feminists seem to endorse empirical success and other traditional empiricist methods and norms for theory choice (Harding, 1993; Rolin, 2006; Wylie, 2001, 2003; Wylie and Hankinson Nelson, 2007). Thus, it is no longer clear what differences there are (if any) between these two views.

My aim is to consider the extent to which there are meaningful differences between contemporary feminist empiricists and standpoint feminists. I will show that recent versions of these views, properly understood, are far more similar than has been assumed by proponents of either view. To the extent that differences exist, I argue that each view needs to adopt features of the other in order to be successful. Thus, feminist empiricists and standpoint feminists should recognize a joint interest in merging as 'feminist standpoint empiricism'.

My intention is not to give a full-scale defense of either view, or of feminist approaches to philosophy of science more generally (though my interpretations will aim to make each view plausible). Rather, I hope to clarify the terrain in feminist philosophy of science and show that feminist standpoint empiricism is more promising than either view on its own. I will begin by examining Harding's characterization of feminist empiricism in order to distinguish it from what I take to be the main tenets of contemporary feminist empiricist accounts.

2. Simplistic feminist empiricism

As stated, one aim of feminist philosophers of science is to identify, understand, and minimize male bias in science. Feminist empiricism, as originally conceived by Harding, was primarily concerned with this aim. According to Harding, feminist empiricism arose as a 'spontaneous consciousness' of feminist researchers in biology and social sciences who began to identify cases of male bias and androcentrism in their respective fields (Harding, 1993: 51). The problem, according to early feminist empiricists, was that scientists were failing to fully adhere to the scientific methods and norms that they themselves accepted as constitutive of good scientific reasoning. Instances of male bias were instances of bad science where scientists allowed gender stereotypes or political values to influence the construction of experiments, the interpretation of data, or the acceptance of theories. The solution, then, was to more rigorously follow existing empiricist norms in order to safeguard objectivity.

Both non-feminist and feminist philosophers of science have criticized this simplistic version of feminist empiricism. Non-feminists have wondered what makes this view 'feminist' in contrast to empiricism in general, as all empiricists would advocate adherence to empiricist methods and norms, and would be equally concerned about bias (Haack, 1998). If feminist empiricism is merely the view that bias is bad and should be prevented by adhering to empiricist norms and methodologies, then it is unclear how it offers any real alternative to traditional empiricism. Early feminist empiricism was also criticized by feminists who maintained it was too conservative, precisely because it endorses problematic tenets of logical positivism. Logical positivists conceived of scientific methods narrowly, obscuring the ways that values and interests influenced scientific reasoning, even when conscientious scientists believed themselves to be following empiricist norms (Longino, 1990).

Yet, contra Harding, contemporary feminist empiricists do not endorse logical positivism (Campbell, 1994). For the sake of this discussion, the feminist empiricists I have in mind include Anderson (1995, 2004), Campbell (1994, 1998), Hankinson Nelson (1990), Kourany (2003), and Longino (1990, 2002), among others. While there are important differences between these authors, they have similarities that make them feminist empiricists. In the next section, I identify what I take to be the main tenets of contemporary feminist empiricism and contrast these with the simplistic conception of empiricism developed by positivists.

3. Contemporary feminist empiricism

Contemporary feminist empiricism is empiricist in that it maintains that scientific theories and models are primarily justified by sense experience. Theories must (i) be empirically successful (or make reliable empirical predictions) when tested with auxiliary hypotheses that do not already assume the truth of the hypothesis being tested, and (ii) meet other cognitive values of theory choice better than alternatives (such as consistency or explanatory power). Cognitive values play an important role to the extent that theories are underdetermined by empirical evidence. When two or more theories are consistent with what has been observed, cognitive values help determine which theories are better justified.

At the same time, contemporary feminist empiricism differs from the empiricism offered by logical positivists in several key ways. First, feminist empiricists have tended to endorse a naturalized approach to understanding science (Anderson, 1995; Campbell, 1998; Hankinson Nelson, 1990). That is, philosophy of science should be informed by empirical studies of scientists producing scientific knowledge. Naturalized epistemology is attractive from a feminist perspective as it allows us to empirically examine and identify the ways in which male bias (as well as other oppressive biases) influences scientific decision-making and practices. This then informs how we should understand and maximize objectivity.

Second, as a result of empirical research into how bias occurs, feminist empiricists argue that the positivist account of observation and empirical testing is too simplistic. Theories can only be tested in conjunction with auxiliary hypotheses or background assumptions, several of which individual scientists are not even aware. Consequently, it is possible for the idiosyncratic values of an individual scientist (including, for example, gender stereotypes or political interests) to operate as background assumptions in testing hypotheses (Longino, 1990: 71–74). Thus, feminist empiricists have argued that background assumptions are best justified when they are carefully and publicly scrutinized by a diverse group of scientists with equal intellectual authority (Hankinson Nelson, 1990; Longino, 1990, 2002). By structuring scientific communities to include scientists with diverse interests and values, who have opportunities to critically assess the methods and reasoning of other scientists, it is more likely that any background assumptions that rest on problematic stereotypes or values will be identified and challenged (Longino, 1990: 73–74, 80; Longino, 2002: 51). Because it is difficult to recognize when one's own values are inappropriately influencing research, a community of scientists with different values will be more likely to detect this influence. On this view, scientific objectivity is located at the level of appropriately structured scientific communities, rather than individual scientists.

Third, unlike logical positivists, feminist empiricists are contextualists with respect to scientific aims, methods, and criteria for theory choice. Feminist empiricists deny that there is one set of aims that characterizes all science. Whether science aims to discover natural laws, explain some phenomena, or generate predictions that will save human lives depends on the particular research context. The aim of generating predictions that will save human lives may exist in biomedicine or climate change research, but not in research contexts that are far removed from human interests and practices (e.g. theoretical physics). Because the aims of science depend on the particular research context, the methods and criteria for theory choice that will promote those aims will also vary. Thus, the justification of scientific theories is contextual in the sense that it occurs within a particular set of assumptions, including aims, cognitive values, and other background assumptions. This contextualism need not led to epistemic relativism (or the abandonment of scientific rationality) so long as the set of aims and assumptions that constitute a scientific research context can be rationally evaluated (Kincaid, 2007; Laudan, 1984; Rolin, 2006; Williams, 2001). This is one reason that feminist empiricists advocate structuring science in ways that promote critical evaluation of aims, methods, and other assumptions.

A fourth difference is that feminist empiricists maintain that in some research contexts, the aims of inquiry can depend upon social and ethical values. While traditional empiricists disagree amongst themselves about what the aims of science are, such as understanding the world (Lacey, 1999), discovering laws of nature, or providing a unifying account of nature (Kitcher, 1993), they have generally held that the aims of science (whatever they are) are always independent of social, political, and ethical aims (Lacey, 1999: 16). Feminist empiricists, however, have argued that the aims of science in particular research contexts can involve social values. For example, Kourany argues that among empirically adequate research programs, we should prioritize those that aim to promote an egalitarian ideal of human flourishing (Kourany, 2003: 6, 10). Similarly, Campbell argues for empiricist methods to be employed towards the aim of 'pursuing the elimination of social bias against women' (Campbell, 1994: 92). Longino has argued that how we understand the epistemic aims of a particular research context can depend on social values. For instance, she argues that when representing scientific phenomena there may be different ways to do so (Longino, 1996: 48). The *best* way to represent something depends on which features we take to be most salient or significant. This, according to Longino, can depend on social as well as epistemic aims. If we want to abolish sexist oppression, then models and theories should 'reveal gender' or use gender as a category of analysis to show how it structures the world (when applicable), as opposed to 'erasing gender' (Longino, 1996: 50).

This is not to say that the social and political aims of science 'trump' the epistemological aims of science. Feminist empiricism does not claim that aims such as empirical adequacy should be subverted to social and political aims for equality. Rather, the idea is that among empirically successful research programs, models, and methods, those promoting equality (or other social aims) should be prioritized or preferred.

This leads to a fifth difference between feminist empiricism and traditional empiricism. Like other empiricists, feminist empiricists acknowledge that other cognitive values (sometimes referred to as epistemic or constitutive values) play a role in helping to determine whether a theory or model is more justified than alternatives. Non-feminist empiricists have traditionally drawn a sharp distinction between cognitive and social values. Cognitive values are thought to include values such as simplicity, consistency, and explanatory power (Kuhn, 1977; Laudan, 2004). These values are *cognitive* insofar as they promote the aims of science (whatever those turn out to be). Because traditional empiricists maintain that the aims of science are independent of social or political values, such values (including feminist values) will be irrelevant to promoting the aims of science and are thus non-cognitive. Feminist empiricists, on the other hand, believe the aims of science are not always independent of social aims and this can also have implications for how we understand cognitive values. For example, Longino (1996) argues that the aim of revealing how gender structures the world may call for valuing complexity, or giving preference to theories or models with greater ontological heterogeneity. Theories and models that posit different gendered entities (as opposed to erasing gender) will be more complex (at least on one interpretation of complexity), but they will better promote the aims of the research context. Longino uses the example of economic models to illustrate this point (1996: 52–53). Models that reveal how gender influences economic transactions are more complex (contain a greater number of economic actors) than simpler models that reduce economic relations to those that occur with 'heads of households'. Traditional 'head of households' models in economics are simpler (have fewer causal actors) but often obscure the economic contributions of women. If the aim of such research involves 'revealing gender', then this supports preferring complex models that allow for the representation of more economic actors.[1]

In this case, social values play an indirect role in helping justify and apply cognitive values. If we value abolishing sexist oppression, this may give rise to a cognitive aim in some research contexts to 'reveal' gender, which may call for preferring complex theories and models to those that are simpler. In other words, the social commitment to equality helps justify our cognitive aims, which in turn justifies cognitive criteria for theory choice. If we had different social values, it might lead to different cognitive aims and different cognitive values (at least in some contexts). Thus, feminist empiricism rejects the view that there is always a sharp distinction between cognitive values and social values. Social values, when they shape the aims of a particular research context, can help justify, interpret, and adjudicate between cognitive values.

At the same time, feminist empiricists do not deny that cognitive values can ground rational theory choice. Longino (1996) suggests that for any particular research context, there will be *some* cognitive criteria (such as empirical success, explanatory power, or ontological heterogeneity) that will help promote the aims of the research. Such values are necessary for members of scientific communities to have shared standards for evaluating theories, methods, and background assumptions. So while cognitive values may not always be independent of social values, cognitive values still play an important role in governing rational theory choice.

A sixth difference between feminist empiricism and traditional empiricism is that feminist empiricism rejects the traditional distinction between 'facts' and 'values'. Some have argued that social values cannot play a legitimate role in science because science is concerned with 'facts'. The idea is that scientific theories are purely descriptive theories about the way the world is. Social and political values, however, are normative and involve judgements about what ought to be the case, or what is ethically or socially desirable. Claims about what *ought* to be the case, it is argued, cannot provide evidence about what *is* the case (Haack, 1998: 127–129).

Feminist empiricism rejects this argument on two grounds. First, some have argued that it is a bit of a red herring (Anderson, 1995). While no one is arguing that it is possible to derive an 'is' from an 'ought', everyone agrees that *some* values (e.g. values about what we take to be good evidence), play a role in scientific theory choice. It is not that we *derive* scientific theories from such values; rather cognitive values help *license* scientific inferences. So, while feminist empiricists make the admittedly controversial claim that cognitive values are sometimes dependent on social values, they are not claiming it is possible to derive an 'is' from an 'ought'.

Second, feminist empiricists have also rejected the fact/value distinction by showing that some scientific hypotheses contain normative concepts. For example, Anderson (2004) argues that sociological research about the extent to which divorce is 'harmful' to children, necessarily rests on value judgements about what constitutes 'harm'. Dupré (2007) has argued that concepts such as 'rape' in animals, 'violence' in children, or 'optimality' in economics are useful scientific concepts, but value-laden nonetheless. Several concepts used in biomedical research such as 'disease' and 'race' have also been shown to rest on social value judgements (De Melo Martín and Intemann, 2007; Intemann, 2000). If there are cases where scientific hypotheses contain normative concepts, it is far less implausible to think that social value judgements play a role as background assumptions in testing such hypotheses.

The seventh difference is that feminist empiricism rejects the traditional distinction between the context of discovery and the context of justification. Logical positivists granted that social and political values might play a role in the context of discovery, or might influence which hypotheses and models were thought of or pursued (Carnap, 1928; Reichenbach, 1938). But they denied that social values had any legitimate role to play in the context of justification, or in determining which hypotheses should be accepted as scientific knowledge. Feminist empiricism argues that such a sharp distinction cannot be made. As stated, feminist empiricism claims that social aims of inquiry can have implications for methodology and criteria for theory choice. In addition, feminist empiricists have pointed out that theory choice is comparative. Scientists must make judgements about whether an empirically successful theory better satisfies cognitive criteria of theory choice than competitors. Yet factors that influence which hypotheses are conceived or developed will influence the range of alternative hypotheses considered in making these comparative evaluations.

A final difference that follows from what has been said above is that feminist empiricism rejects the value-free conception of scientific objectivity and rationality. Because social and political values can play a legitimate role in shaping the aims of research, helping to justify and adjudicate between cognitive values, and operating as background assumptions in testing theories with normative content, feminist empiricism maintains that 'objective' or 'rational' science is not necessarily free of such values. Feminist empiricism's social account of objectivity does not require the elimination of social values, but attempts to make such values explicit so that they can be critically evaluated and debated along with other background assumptions. Thus, objectivity is taken to emerge not from the value-neutrality of individual scientists, but from a neutrality of the procedures built into the structure of scientific communities to identify and critically evaluate normative assumptions.

To summarize, eight key differences distinguish contemporary feminist empiricism from the positivist version of empiricism that underlies Harding's caricature:

1. Taking a naturalized approach to understanding science.
2. Locating justification and objectivity at the level of communities, rather than individuals.
3. Endorsing contextualism about aims, methods, and theory justification.
4. Asserting that the aims of scientific research, properly understood, can and sometimes should include social and political aims, including feminist aims.
5. Rejecting the traditional cognitive/non-cognitive value distinction, but retaining shared criteria for theory choice (including empirical adequacy).
6. Rejecting the traditional distinction between context of discovery and justification.
7. Rejecting the traditional fact/value distinction.
8. Adopting a procedural conception of objectivity that is not 'value-free'.

Of course, many non-feminist contemporary empiricists also endorse some of these features. Yet non-feminist versions of empiricism do not have all of the features, particularly number four. While feminist empiricism is *empiricist*, it is also *feminist* insofar as it is committed to advancing certain feminist aims and goals. In particular, feminist empiricism is thought to be able to advance the aims of (1) reducing gender, race, and class bias in science, (2) explaining why diversity, or the participation of historically underrepresented groups, is important in science, (3) establishing a positive role for feminist values in science, and (4) making possible as well as prioritizing the kind of research that is needed to promote egalitarianism.

As stated at the outset, my aim is not to defend feminist empiricism as a philosophy of science. As a *feminist* philosophy of science, however, it has been taken to be competing with standpoint feminism. In the next section, I examine the main theses of standpoint feminism and compare these with feminist empiricism. While standpoint theorists may be guilty of arguing against a 'straw person' of feminist empiricism, feminist empiricists are also guilty of misinterpreting the central claims of standpoint feminists. I will show that the main tenets of standpoint theory are far closer to those of feminist empiricism than proponents of either view have been willing to admit.

4. Standpoint feminism

Standpoint feminism arose in the 1970s and 1980s (Harding, 1986; Hartsock, 1983; Hill Collins, 1991; Rose, 1987; Smith, 1974) and continues to be controversial to both feminist, as well as non-feminist, philosophers of science. It consists of three main theses:

1. *The Situated Knowledge Thesis*: Social location systematically influences our experiences, shaping and limiting what we know, such that knowledge is achieved from a particular standpoint (adapted from Wylie, 2003).
2. *The Thesis of Epistemic Advantage*: Some standpoints, specifically the standpoints of marginalized or oppressed groups, are epistemically advantaged (at least in some contexts) (Wylie, 2003: 28).
3. *The Methodological Thesis*: Research that involves or affects marginalized groups should begin with the lives and experiences of those marginalized groups. In other words, researchers should engage in 'studying up', or examining power relations, institutions, policies,

and technologies that perpetuate oppression from the perspective of the oppressed (adapted from Crasnow, 2006: 835).

Both feminist empiricists and non-feminists have routinely interpreted these theses in ways that make them seem either obviously false or trivially true. Standpoint feminism has been conflated with the view that women have a distinct way of knowing different from men. Interpreted in this way, standpoint feminism has been charged with reinforcing gender stereotypes and falsely assuming that all women have some shared standpoint (Bar On, 1993; Haack, 1998; Hekman, 1997).

Standpoint feminists have also been understood as claiming that women have an automatic epistemic privilege in virtue of being oppressed (Haack, 1998; Hekman, 1997; Pinnick et al., 2003). Membership in an oppressed group is taken to be sufficient for having a less distorted view of the world. This seems false as we can easily think of cases where members of oppressed groups actually have a less accurate view of the world (such as cases where oppressed individuals have internalized their oppression).

Some (e.g. Longino, 1993) have argued that the first two theses are inconsistent. If some standpoints are epistemically advantaged, then there must be some standpoint-neutral or 'objective' perspective to judge which standpoints are epistemically better. But, then, it cannot be that all knowledge is situated to a particular standpoint, otherwise no such objective standard would exist (Longino, 1993: 106–107).

Others have interpreted the theses of standpoint feminism in ways that make them more plausible, but far less interesting. For example, one might interpret the situated knowledge thesis as merely claiming that people have different experiences and so may have different knowledge. Women who give birth to children, for instance, have experiences that others do not. Their experiences give them knowledge about what it's like to give birth, or how the pain of childbirth compares to other painful experiences. Most would agree that while this is true, it does not have any interesting implications for philosophy of science. Presumably differences in these kinds of experiences would only apply to a small range of cases that are irrelevant to the majority of scientific inquiry.

Thus, standpoint feminism has been interpreted in ways that make it implausible or irrelevant. But, as with feminist empiricism, it is unclear that contemporary standpoint feminists hold the caricaturized views attributed to them.[2] Recent work has clarified the core insights of earlier standpoint theorists in ways that reject the simplistic interpretations described above (Crasnow, 2006; Harding, 2004; Hartsock, 1997; Rolin,

2006; Wylie, 2003). In particular, Wylie (2003: 28) has argued that for standpoint theory to be viable it must not presuppose an essentialist definition of the social categories by which standpoints are characterized, and it must not maintain that standpoints of the oppressed are automatically epistemically advantaged. I will examine recent accounts of the three standpoint theses and consider whether the more sophisticated interpretations are substantially different from feminist empiricism.

The situated knowledge thesis

The situated knowledge thesis can be broken into two parts: (a) the claim that social location systematically influences our experiences, shaping and limiting what we know, and (b) that knowledge is achieved from a particular standpoint. Each of these two components will be considered in turn.

Standpoint feminism argues that, historically, systems of oppression (such as racism, sexism, heterosexism, and classism) have influenced the material circumstances of individuals (such as their living conditions, their opportunities, and their treatment in a variety of social situations). As a result, individuals from different social locations have, to some extent, different experiences. The ways that social location shapes experience are not homogeneous within a particular social group. Women, for example, will have different experiences in virtue of their other differences such as race, class, geographical location, etc. Nonetheless, certain social features are relevant to the kinds of experiences one has because of how they affect one's material circumstances.

On this view, the extent to which particular aspects of social location shape experience is a contingent matter. If, for example, sexism were to be abolished, then sex would not shape our experiences in the same ways. Thus, the features of one's social location that have epistemic significance can change over time. Standpoint feminism maintains that at this time sex, race, class, sexual orientation, citizenship, and other currently marginalized social categories can significantly shape experiences.

Standpoint feminists have identified a variety of ways that experiences shaped by social location have epistemic impacts. Some standpoint theorists have argued that experiences and interests related to geographical location and class can influence how research questions are framed (Shiva, 2006). Research on genetically modified food, for example, has often been framed as 'How can we maximize the production of a crop under certain (optimal) material conditions?' rather than 'How can we produce crops so that all the people in a given region will gain access to a well-balanced diet in a context that enhances local agency and sustains the environment?' (Lacey, 1999: 189–196; Shiva, 2006: 240). Researchers with different life experiences are likely to pose alternative research questions and view the aims of the research differently.

Standpoint feminism also maintains that experiences of researchers (informed by their social location) can influence which background assumptions seem reasonable in designing experiments, deciding what to count as evidence, and interpreting data. Fausto-Sterling (1985: 32) illustrates how this occurred in research on sex differences in visual-spatial ability. Male researchers failed to see flaws with certain experiments because they lacked the kind of experiences that would have led them to identify problematic assumptions (such as the fact that some experimental tests were inadvertently designed in ways that hindered female performance).

Homogeneous experiences may also limit the range of alternative hypotheses considered. In visual-spatial research, male researchers did not consider whether the experimental results might be explained by differences in the socialization of males and females (Fausto-Sterling, 1985: 32). This is not to say that it would have been impossible for a male scientist to think of these alternative explanations. But the extent to which scientists are able to think of alternative hypotheses is partly influenced by their experiences and capacity for creative thinking. It is *easier* to see the flaws in experiments when one has had experiences that contradict the underlying assumptions.

Standpoint feminism also argues that the social location of researchers can influence what is taken to be evidence. Wylie argues that as women entered the field of archaeology in the 1980s, their experiences led them to challenge assumptions about what counted as 'data' and opened up new lines of evidence (Wylie, 2001: 25; Wylie and Hankinson Nelson, 2007: 63–65). In particular, male researchers appeared to have ignored several artifacts that were used in subsistence activities associated with females (Gero, 1991; Wylie and Hankinson Nelson, 2007). Female archaeologists challenged assumptions about what counted as a 'tool' and discovered new evidence about the evolution of tool use in humans.

Thus, the first part of the situated knowledge thesis amounts to the claim that social location can influence how research questions are framed, the selection of background assumptions (including those used to interpret observations), and the pool of alternative hypotheses considered. To a large extent, this view is compatible with feminist empiricism. Like feminist empiricism, standpoint feminism appeals to experience as playing a key role in the justification of beliefs. Standpoint feminists similarly take a naturalized approach to examine how the social make-up of scientists influences scientific inquiry. The key difference worth noting here is that standpoint feminists make the additional claim that *social location* in particular can influence the experiences one has. One important implication of this view is that knowledge (or at the very least experience) is embodied. That is, not all bodies have the same experiences and knowledge acquisition cannot be understood merely in terms of interactions between a universal (sexless, raceless, classless) sensory apparatus and a rational mind. This helps explain why the historical absence of women and other marginalized groups from many science disciplines is epistemically (not just morally) problematic. Feminist empiricists, in contrast, have traditionally focused on how individuals with diverse *values* can promote the objectivity of scientific communities, but have not posited a special role to the *social location* of researchers *per se*. I later examine the implications of this difference in further detail. First, however, I will turn to the other part of the situated knowledge thesis.

The second part of the situated knowledge thesis claims all knowledge is achieved from a particular standpoint. What this claim amounts to depends on how we understand the concept of a 'standpoint'. It is tempting to think of a standpoint as a social location, or a perspective. Yet standpoint theorists have taken pains to distinguish a 'standpoint' from a 'perspective'. There are, I believe, three main characteristics of standpoints that distinguish them from socially located perspectives. First, standpoint feminists have stressed that a standpoint is *achieved*, rather than something that occurs automatically in virtue of one's social location (Harding, 2004; Rolin, 2006; Wylie, 2003). What exactly is achieved? Standpoint feminism claims that the achievement of a standpoint involves a critical conscious reflection on the ways in which power structures and resulting social locations influence knowledge production. For Harding, a standpoint is a distinctive insight about how hierarchical social structures work (Harding, 2004: 31). Wylie states that a standpoint is 'a critical consciousness about the nature of our social location and the difference it makes epistemically' (Wylie, 2003: 31).

This critical consciousness, however, is not something that is achieved by individuals. A second feature that distinguishes a standpoint from a perspective is that it is something achieved by *communities*, rather than individuals. Consciousness-raising groups of the 1970s provide an example of how groups of women were able to reflect on their individual experiences and collectively identify patterns they were not able to recognize on their own (Smith, 1992). Individual women had experiences, such as being groped by a male co-worker, which they had previously interpreted as accidental, imagined, or deserved. But, when multiple women in the group reported similar experiences, their individual interpretations became inconsistent with the data. As a group, women were able to see patterns in their experiences, identify relationships between those patterns and oppressive arrangements, and achieve an understanding of how systems of oppression limit and shape their knowledge.

This brings us to the third characteristic that distinguishes a standpoint from a perspective. A standpoint is explicitly *committed to certain moral and social values*. According to Harding, a standpoint 'intends to map the practices of power, the ways the dominant institutions and their conceptual frameworks create and maintain oppressive social relations' (2004: 31). Adopting a feminist standpoint involves making a normative commitment to revealing the ways in which gender, for example, shapes and limits scientific inquiry as well as what we take to be scientific knowledge. Thus, some have argued that standpoints require a kind of activism (Hundleby, 1997). They require understanding and revising our epistemic practices so as to identify, understand, and ultimately abolish, the ways in which systems of oppression limit knowledge production.

This interpretation of 'standpoint' helps clarify the claim that knowledge is achieved from a standpoint. Knowledge is best achieved by groups who share a set of normative commitments about the aims of inquiry and who allow for conscious and critical interactive evaluation of hypotheses, background assumptions, interpretations of evidence, and methods. The social locations of individual group members, insofar as they produce different experiences relevant to the critical evaluation of background assumptions, can make a difference to how epistemically rigorous this critical reflection is. As Rolin argues:

> [O]pening a community to wider participation as well as to outside criticism increases the likelihood that some default assumptions are challenged in appropriate ways. The more diversity there is in a scientific community, the more likely it is that its default assumptions are challenged, and consequently either defended, modified, or abandoned. So, I suggest that a standpoint is a commitment to diversity in a scientific community. (Rolin, 2006: 135)

On this interpretation, however, the situated knowledge thesis begins to be indistinguishable from tenets of feminist empiricism. On both views, knowledge is achieved by communities rather than by individuals because it is difficult for individuals to recognize the ways in which their own experiences and values may be influencing their reasoning. But, by establishing diverse epistemic communities that critically reflect on the assumptions made by individual scientists, it is more likely that any problematic assumptions will be recognized, challenged, evaluated, and revised. Both views also share a commitment to structuring inquiry in ways that help identify and challenge systems of oppression in epistemic practices and in the world. Towards this end, both views endorse a naturalized approach to scientific inquiry. Each view requires us to examine how systems of oppression limit and shape what we know, and to use these empirical investigations to inform how we structure scientific practices. Finally, both views take diversity within scientific communities to be crucial in achieving maximal critical assessment of background assumptions operating at every level of scientific inquiry. Though, as we have seen (and as will be discussed in more detail later), there may be differences in the kind of diversity that promotes these epistemically desirable effects.

The thesis of epistemic advantage

I will now turn to standpoint feminism's thesis of epistemic advantage, which is arguably its most controversial claim. Many feminist empiricists deny this thesis and take this to be a key difference between the two views. Again, a rough statement of the thesis is that some standpoints, specifically the standpoints of marginalized or oppressed groups, are epistemically advantaged (at least in some contexts) (Wylie, 2003: 28). In order to understand this thesis fully, it is important to identify the nature of the advantage (what it consists in), its source (why or how it accrues), and its scope.

Standpoint theory argues that members of oppressed groups have some epistemic advantages in virtue of their special location as 'insiders-outsiders' (Hill Collins, 1991). Members of oppressed groups must understand the assumptions that constitute the worldviews of dominant groups, in order to successfully navigate the world. At the same time, they often have experiences that conflict with dominant views and generate alternative views about how the world works (Wylie, 2003: 34–35). Hill Collins uses the example of a black female sociologist. On the one hand, in order to be successful as a sociologist, she must learn and conform to what is taken to be authoritative knowledge in the field. Yet to the extent that she has had experiences as a black woman that conflict with these assumptions, it reveals the extent to which widely held assumptions are in fact unjustified or partial (Hill Collins, 1991: 49–51).

To understand the potential benefits of the insider-outsider perspective in science, consider 1950s research on racial differences in impulse control. One famous study conducted by Walter Mischel aimed to determine whether there were racial differences in impulse control stemming from differences in the ability to evaluate the long-term consequences of actions (Mischel, 1958).[3] The experiment involved two groups of school children in Trinidad: one group of black students and one group of non-black students. A white experimenter presented each child with the choice of either having one small piece of candy immediately, or returning the next day and receiving a much larger piece of candy (Mischel, 1958). Black children tended to take the immediate piece of candy, while the non-black children overwhelmingly opted to wait until the next day to receive more candy. Mischel concluded: 'Negroes are impulsive, indulge themselves, settle for next to nothing if they can get it right away, do not wait for bigger things in the future but, instead, prefer smaller things immediately' (Mischel, 1958: 59).

Of course, looking at this case in retrospect, Mischel's reasoning and methodology seem outrageous. There were more plausible alternative hypotheses that could have accounted for the differences in decision-making along racial lines. In particular, black students were more likely to be distrustful of white scientists. Thus, it was more rational for black students to take 'what they could get' immediately, as opposed to risk getting nothing later. Or, perhaps the differences in decision-making reflected socio-economic differences. Insofar as black students tended to be poor (and thus more likely to be starving), delaying gratification may have been more difficult. But the researchers involved in this experiment did not consider these alternative explanations. It would have been difficult for them to identify their own problematic background assumptions because it is unlikely they had the kind of life experiences that would have made them aware that black children may have good reason to be suspicious and distrustful of white scientists, or that some of these children may not have eaten and be concerned about their next meal. Again, this is not to say that it would have been *impossible* for white scientists to think of these alternative explanations. But, because of their experiences as 'outsiders', black researchers were more likely to know that members of the black community are distrustful of whites (especially those promising delayed rewards), particularly as this is often not told to whites because of the potential repercussions of admitting it. At the same time, as 'insiders' black researchers had the training to understand and identify how problematic assumptions were operating in the research and might have proposed alternative experiment designs, hypotheses, or interpretations of data to correct potential bias. In this way, members of oppressed groups can have epistemic advantages when they have experiences and interests that (i) give them access to a wider range of empirical evidence, (ii) allow them to identify alternative hypotheses that are not likely to occur to others, (iii) identify problematic background assumptions, and (iv) allow them to see ways in which systems of oppression influence knowledge production.

These potential epistemic advantages are not automatic for several reasons. First, it is not sufficient simply to be a member of an oppressed group because, as stated earlier, one must be able to critically reflect on the ways in which systems of oppression have shaped those experiences. Second, members of oppressed groups may also face obstacles to being 'insiders' that may hinder any epistemic advantages their experiences might yield. Wylie points out that members of oppressed groups often lack access to other key epistemic resources, such as 'certain kinds of information, the analytic skills acquired through formal education, and a range of theoretical and explanatory tools' (Wylie, 2003: 37). Moreover, potential epistemic advantages for members of oppressed groups can only yield new knowledge if they are empowered to have their criticisms listened to and taken seriously. Finally, members of oppressed groups will not have a 'blanket' epistemic advantage. That is, there may be a range of areas where their experiences are not relevant to the phenomenon being studied or how it is studied, such that members of oppressed groups will not necessarily have any special insight. (Though, it is perhaps worth noting that it is difficult to predict a priori when such experiences might be relevant.) Thus, whether members of oppressed groups have epistemic advantages, as well as the scope of those advantages, depends on several factors.

Yet, so far, my discussion makes it sound like epistemic advantage accrues to *individuals*. This, I believe, reveals an important ambiguity in standpoint feminism. Many of the examples that standpoint feminists have given of epistemic advantage are of advantages that individuals possess. Yet the thesis of epistemic advantage claims that it is certain *standpoints* that have an epistemic advantage and, as we have seen, standpoint feminists also maintain that standpoints are achieved by *groups* and not by individuals. How are these claims to be reconciled?

I submit that the thesis of epistemic advantage can best be interpreted as the claim that epistemic communities that are socially diverse and include members of historically underrepresented groups have certain epistemic advantages (at least in some contexts). Members of oppressed groups can play an important role in scientific or other epistemic communities when they have experiences that are relevant to the research and allow them to suggest alternative ways to frame research questions, generate alternative or new hypotheses, and challenge existing background assumptions that have been taken for granted.

Understood in this way, the thesis of epistemic advantage seems similar to claims made by feminist empiricism. As we have seen, Longino argues that diverse scientific communities are more likely to rigorously scrutinize background assumptions, propose alternative hypotheses, or reframe research questions. Both views thus take diversity to promote objective scientific communities. Again, the remaining difference is that standpoint feminists argue that diversity of *experiences* or social location is particularly important in this regard. Before considering this difference in more detail, I will briefly turn to the last thesis of standpoint theory.

The methodological thesis

The methodological thesis claims that research involving or affecting marginalized groups should begin with the lives and experiences of those marginalized groups. In other words, researchers should engage in 'studying up', or examining power relations, institutions, policies, and technologies that perpetuate oppression from the perspective of the oppressed (adapted from Crasnow, 2006). Some might take the methodological thesis to be an implication of the thesis of epistemic advantage. If members of oppressed groups have certain epistemic advantages, then we should start with them. I have argued, however, that the thesis of epistemic advantage is best understood as a claim about epistemic communities, rather than individuals. Thus, I take the methodological thesis to be a distinct thesis about how certain phenomena should be studied.

The methodological thesis may be better seen as a consequence of standpoint feminism's normative commitment to understand and abolish systems of oppression. There are multiple ways to study a particular phenomenon, and some of these will be more useful than others depending on the social aims of the research. Consider, for example, research measuring climate change impacts. Studies on the impacts of climate change must make assumptions about what will count as an 'impact' and how, exactly, impacts should be measured. Climate change impact studies often rely solely on cost-benefit analyses to compare the 'harmful' effects of various climate change adaptation or mitigation strategies (Schneider et al., 2000). Cost-benefit models have two potential drawbacks. First, such models typically only assess damages that can be easily quantified in monetary terms by their market value, such as the number of lives lost, the cost of property destruction, and other economic variables. They have often neglected other 'harmful impacts' that are not as easily quantified such as impacts to quality of life, cultural traditions, and biodiversity. Second, such models only measure the overall cost-benefit ratio of various scenarios and do not provide information about the distribution of harms and benefits (Schneider et al., 2000). Thus, such models will not provide information about which mitigation or adaptation strategies will achieve a socially just or equitable distribution of benefits and burdens in dealing with climate change. This has led some (Schneider et al., 2000) to adopt alternative models that can better measure a wider range of potential harms, as well as their distribution.

'Studying up' requires researchers to approach the problem of climate impacts from the lives and experiences of those who are marginalized and most affected by climate change (i.e. poor people living in developing counties in the southern hemisphere). Citizens of developing counties, for instance, may be more concerned with knowing how the benefits and burdens of mitigation (or non-mitigation) will be distributed. From their perspective, impacts to farmlands or loss of culture may be just as devastating as loss of life. Their experiences may help researchers identify problematic assumptions in existing theories and models that conflict with the material conditions 'on the ground' in areas affected by the research. Thus, if the aim of such research is not just to produce true beliefs about climate change, but true beliefs that also address the social aim of preventing the greatest impacts to the most vulnerable populations, or of generating climate policy solutions that help prevent, rather than perpetuate, further socio-economic inequalities, 'studying up' provides a more adequate methodology.

To what extent does the methodological thesis illustrate a substantive difference between standpoint theory and feminist empiricism? On the one hand, feminist empiricism is open to the possibility of employing methods that are empirically successful towards promoting the aims of research. Furthermore, as we have seen, feminist empiricism agrees that certain social and political aims are often bound up with the epistemic aims of science. Feminist empiricists have urged prioritizing scientific inquiry that works towards egalitarian aims. 'Studying up', then, insofar as there is evidence that it is an empirically successful methodology in helping promote the epistemic and social aims of science, is at least *consistent* with feminist empiricism. Yet historically, feminist empiricists have not explicitly endorsed the methodological thesis.

To recap, there are several key similarities between feminist empiricism and standpoint feminism, as well as a few key differences:

Sophisticated feminist empiricism	Standpoint feminism
1. Takes a naturalized approach to understanding science.	1. Same.
2. Locates justification and objectivity at the level of communities, rather than individuals.	2. Same.
3. Endorses contextualism about aims, methods, and theory justification.	3. Same.
4. Asserts that the aims of scientific research, properly understood, can and sometimes should include social and political aims, including feminist aims.	4. Asserts that liberatory standpoints involve a normative commitment to expose and challenge systems of domination and subordination.
5. Rejects the traditional cognitive/non-cognitive value distinction, but retains shared criteria for theory choice (including empirical adequacy).	5. Rejects the traditional cognitive/non-cognitive value distinction and does not explicitly endorse empiricist criteria for theory choice.
6. Rejects the traditional distinction between context of discovery and justification.	6. Same.
7. Rejects the traditional fact/value distinction.	7. Same.
8. Adopts a procedural conception of objectivity that is not 'value-free'	8. Same.
9. Maintains that diversity of values and interests within epistemic communities is necessary for increasing objectivity.	9. Maintains that diversity of social locations or experiences within epistemic communities is necessary for increasing objectivity.
10. Does not explicitly endorse 'studying up'.	10. Is committed to 'studying up' from the lives and experiences of oppressed groups when such groups are affected by the research.

Both views, at least as they are being articulated by contemporary advocates, have a great deal in common. Numbers 1, 2, 3, 6, 7, and 8 are identical. Number 4, although expressed slightly differently, I take to be essentially the same idea. That is, both views maintain that some research should aim to expose and challenge oppressive systems. This is why Longino (1996) urges research to 'reveal gender' and Kourany (2003) urges research that promotes egalitarianism. Neither view would claim that this is the only sort of research that should be done, but that there is reason to prioritize such research. Similarly, feminist empiricism is at least consistent with standpoint feminism with respect to 10. Feminist empiricism would be committed to 'studying up' insofar as there was good empirical evidence that this promotes either epistemic or egalitarian aims that it shares with standpoint feminism.

The main remaining differences between the two views can be seen in 5 and 9. Even with respect to these differences, it is not clear that the disagreement is substantive. Wylie (2003), for example, has explicitly adopted a similar framework to feminist empiricists in thinking about cognitive values (number 5 above). And Campbell (1994) has argued that feminist empiricism can, in principle, adopt the standpoint's position on 9. He argues that feminist empiricism can incorporate insights about how individuals from certain social locations might have epistemic advantages insofar as this is supported by empirical evidence.

I want to make the stronger, claim, however, that standpoint feminism, in order to be successful, must adopt the feminist empiricist position with respect to 5. At the same time, feminist empiricists should adopt standpoint insights with respect to 9 in order to avoid certain problems with their view. In short, my thesis is that the two views should be merged and that the resulting view, 'feminist standpoint empiricism', will be stronger than either view alone.

5. Why the two views need each other: towards feminist standpoint empiricism

As we have seen, standpoint feminism does not explicitly endorse an empiricist view of theory justification. Feminist empiricism maintains that theories are evaluated in terms of their empirical success (or ability to make successful empirical predictions in conjunction with background assumptions), along with the extent to which they better meet cognitive values than competitors. Even for feminist empiricists, cognitive values are not necessarily fixed or absolute. Cognitive values, as well as how they are weighed in relation to each other, can depend on the particular aims of the research context. Feminist empiricists take a naturalized approach towards identifying and understanding why certain criteria have produced better (or worse) theories in the history of science. This has allowed them to acknowledge the virtue of many traditional cognitive values such as explanatory power, consistency, simplicity, and fruitfulness. At the same time, feminist empiricists have been able to show that there are contexts where other non-traditional criteria for theory choice (such as ontological heterogeneity and novelty) can also be valuable (Longino, 1996). These alternative criteria are justified insofar as they promote the aims of at least some research contexts.

Although standpoint feminism has not explicitly endorsed an empiricist conception of cognitive values and theory choice, there are several reasons for it to do so. First, such an account can explain successes in the history of science that do indeed seem to be guided by traditional criteria for theory choice. This also provides common ground for being able to engage non-feminist philosophers of science and many scientists. Feminist philosophy of science has a better chance of gaining proponents when it can point out that considerations accepted by traditional empiricists better justify feminist alternatives.

There is, however, a second stronger reason why standpoint feminists should adopt an empiricist framework for theory justification. Standpoint feminists must grant the importance of empirical success and other cognitive values to maintain the coherence of their own view. The situated knowledge thesis maintains that social location helps shape individual experiences. This is only epistemically significant insofar as experiences help justify our beliefs. Moreover, because individuals can have *different* experiences, and hence access to different evidence, individuals within a standpoint must have some shared values in order to, as a community, critically assess those experiences and come to a consensus about what to believe. That is, members of a standpoint must have some shared cognitive values, otherwise they will be unable to engage in critical discussions about their experiences or arrive at the kind of joint reflective consciousness that a standpoint requires. In order for critical discussions (or any meaningful communication) to occur, there must be some shared norms to govern the evaluations of claims and experiences. There must be some shared standards for what counts as reason or evidence. Thus, standpoint feminists must endorse the feminist empiricist position about cognitive values because without shared standards for a particular research context a standpoint of any kind would not be possible. To be sure, particular standpoint feminists might take issue with particular cognitive values that have traditionally been endorsed by empiricists. But, there is already disagreement among feminist empiricists (and among empiricists more generally for that matter) about which particular values should be endorsed as cognitive. For example, many have disagreed about whether simplicity should be valued over complexity. So, in this regard, I believe there is just as much debate amongst feminist empiricists as between them and standpoint feminists. My point is that standpoint feminists must acknowledge that there are some shared cognitive values that govern a particular standpoint.

Yet while standpoint feminism would benefit from the resources of empiricism, so too would feminist empiricism benefit from incorporating standpoint's account of diversity within scientific communities. Both views take diversity to be instrumental in increasing objectivity of scientific communities by generating new lines of inquiry, expanding the range of alternative hypotheses considered, and challenging problematic background assumptions that are taken for granted. But feminist empiricism has traditionally maintained that objectivity is promoted by achieving research communities that are comprised of those with diverse *values and interests*, whereas standpoint feminism has argued for the importance of diversity of *social locations or experiences*. Which kind of diversity better promotes objectivity?

The feminist empiricist account of diversity is ultimately problematic. On this account, all values are treated as equally desirable for contributing to diversity. That is, sexist, racist, and other questionable values would be just as important to achieving diversity within an epistemic community as feminist ones. Creationist values appear to be just as important to research in biology as ones held by evolutionary theorists precisely because they would help achieve diversity of values and interests. This is counterintuitive, but poses a deeper dilemma for feminist empiricism. The question is: Are some values and interests more justified than others or not? If some values are more justified than others, then it is unclear why a diversity of values is useful for achieving rational critical evaluation of research programs, hypotheses, and background assumptions. Creationists might challenge assumptions about carbon dating, and racists might argue that any interpretation of data that is inconsistent with other beliefs about white superiority must be wrong. But if these challenges presuppose values that are themselves unjustified, then it is unclear why they should be taken seriously. On the other hand, if feminist empiricism maintains that value judgements are not themselves justified (and that is why scientific communities must include representatives of diverse values), then this is inconsistent with feminist empiricism's own normative commitment to promoting feminist political and social aims. That is, if feminist empiricism denies that some values are more justified than others, they deprive themselves of the resources they need to justify *feminist* aims and values.

Standpoint feminism provides an alternative account of diversity that does not lead to this tension. Standpoint feminists claim that scientific communities should consist of diverse individuals from different social locations because they maintain that social location shapes individual experiences. In this way, they maintain that knowledge is embodied.

Different bodies (in virtue of differences in biology and differences in the ways those bodies are treated or causally interact with power structures, social institutions, and other material circumstances) lead to different experiences. Since experience justifies beliefs, it is clear why diversity of experience is important: it gives the community as a whole access to more of the relevant evidence for critically evaluating hypotheses and assumptions. Standpoint theory can also explain why different kinds of social diversity may be important to different research contexts. For example, in climate impacts research, geographical origin is likely to track differences in experience that may bear on the research. In this case, sexual orientation may have less relevance. Thus, standpoint feminists can maintain that the kind of social diversity needed in any particular context will partly depend on which experiences are most relevant to the research. Because standpoint feminism is not committed to a diversity of values within a standpoint, they can consistently argue that some values are more justified than others. Thus, standpoint feminism can consistently adopt a normative commitment to identifying and challenging systems of oppression, as it can maintain that these aims are better justified than alternatives.

In conclusion, both feminism empiricism and standpoint feminism could be made stronger by incorporating features of the other view. Standpoint feminists need to adopt an empiricist account of theory justification (including the empiricist framework for cognitive values) in order to make sense of how standpoints are achieved and are epistemically successful. Feminist empiricism should acknowledge the ways in which social location shapes experience (as this appears to be empirically supported) and adopt the standpoint account of diversity within ideal scientific communities, as this is more consistent with their own normative commitments.

Yet in adopting these features of each other's accounts, the two views would become indistinguishable. To be sure, there may be remaining differences between some individual proponents of feminist empiricism and standpoint feminism. Yet these differences are no greater than differences among, for example, feminist empiricists. Potential problems with both views can best be addressed, then, by reconfiguring the terrain in feminist philosophy of science and moving towards a common feminist standpoint empiricism. Such a view would adopt the strengths of each view while avoiding problems that each view has on its own.

As admitted at the outset, I have not responded to all of the possible objections that these views face (individually or combined). In particular, some will be skeptical of the claim, central to most feminist philosophies of science, that social values can play a role in good science. Skeptics have argued that such values only serve to 'politicize' science in ways that interfere with scientific rationality. Numerous arguments have been given to show that the incorporation of social and political values need not undermine scientific objectivity and rationality; arguments made by both feminists (Anderson, 2004; Crasnow, 2006; Longino, 2002; Wylie and Hankinson Nelson, 2007) and non-feminist empiricists (Kincaid, 2007; Kitcher, 1993; Lacey, 1999; Solomon, 2001). I will not repeat those here. I will point out, however, that making this kind of argument is a task that also unites feminist empiricists and standpoint feminists. Under feminist standpoint empiricism, feminist philosophers of science can stop debating old theoretical distinctions that no longer apply and pursue a common agenda of empirical research into how the experiences of marginalized groups can contribute to better science.

Notes

1. This is not to say that social values (such as promoting equality) give us reason to believe that certain theories or models are more likely to be true. Rather, the example shows that certain criteria for theory choice (e.g. complexity) better promote the other cognitive aims of science (such as revealing gender), which are partly determined by our social values (e.g. ending sexist oppression). Thus, social values can give rise to particular cognitive aims, which then help determine cognitive criteria for theory choice.
2. In fact, several have argued that many of the criticisms of standpoint feminism are based on misinterpretations or misreadings of even early standpoint feminists; see Crasnow (2006); Hartsock (1997); Hill Collins (1997); Smith 1997; Wylie 2003.
3. For more discussion of this example, see Miller (1987: 173–177).

References

Anderson, E. (1995) 'Feminist Epistemology: an Interpretation and a Defense', *Hypatia*, 10(3): 50–84.

Anderson, E. (2004) 'Uses of Value Judgments in Science: a General Argument, with Lessons from a Case Study of Feminist Research on Divorce', *Hypatia*, 19(1): 1–24.

Bar On, B. (1993) 'Marginality and Epistemic Privilege', in L. Alcoff and E. Potter (eds), *Feminist Epistemologies* (New York: Routledge), 83–100.

Campbell, R. (1994) 'The Virtues of Feminist Empiricism', *Hypatia*, 9(1): 90–115.

Campbell, R. (1998) *Illusions of Paradox: a Feminist Epistemology Naturalized* (Lanham, MD: Rowman & Littlefield).

Carnap, R. (1928/1967) *The Logical Structure of the World and Pseudoproblems in Philosophy* (Berkeley: University of California Press).

Crasnow, S. (2006) 'Feminist Anthropology and Sociology: Issues for Social Science', in S. Turner and M. Risjord (eds), *Handbook of the Philosophy of Science*, vol. XV, *Philosophy of Anthropology and Sociology* (Amsterdam: Elsevier).

De Melo-Martín, I. and K. Intemann (2007) 'Can Ethical Reasoning Contribute to Better Epidemiology? A Case Study in Research on Racial Health Disparities', *European Journal of Epidemiology*, 22: 215–221.

Dupré, J. (2007) 'Fact and Value', in H. Kincaid, J. Dupré, and A. Wylie (eds), *Value-Free Science? Ideals and Illusions* (Oxford: Oxford University Press), 27–42.

Fausto-Sterling, A. (1985) *Myths of Gender: Biological Theories About Women and Men* (New York: Basic Books).

Gero, J. M. (1991) 'Genderlithics: Women's Roles in Stone Tool Production', in J. M. Gero and M. W. Conkey (eds), *Engendering Archaeology: Women and Prehistory* (Oxford: Blackwell).

Haack, S. (1998) *Manifesto of a Passionate Moderate* (Chicago: University of Chicago Press).

Hankinson Nelson, L. (1990) *Who Knows? From Quine to a Feminist Empiricism* (Philadelphia: Temple University Press).

Harding, S. (1986) *The Science Question in Feminism* (Ithaca, NY: Cornell University Press).

Harding, S. (1991) *Whose Science? Whose Knowledge? Thinking from Women's Lives* (Ithaca: Cornell University Press).

Harding, S. (1993) 'Rethinking Standpoint Epistemology: What is Strong Objectivity?', in L. Alcoff and E. Potter (eds), *Feminist Epistemologies* (New York: Routledge), 49–82.

Harding, S. (2004) 'A Socially Relevant Philosophy of Science? Resources from Standpoint Theory's Controversiality', *Hypatia*, 19(1): 25–47.

Hartsock, N. (1983) 'The Feminist Standpoint: Developing the Ground for a Specifically Feminist Historical Materialism', reprinted in S. Harding (ed.), *The Feminist Standpoint Theory Reader* (New York: Routledge), 35–54.

Hartsock, N. (1997) 'Comments on Hekman's "Truth and Method": Truth or Justice?', *Signs*, 22(21): 367–374.

Hekman, S. (1997) 'Truth and Method: Feminist Standpoint Theory Revisited', *Signs*, 22(21): 341–365.

Hill Collins, P. (1991) 'Learning from the Outsider Within', in M. M. Fonow and J. Cook (eds), *Beyond Methodology: Feminist Scholarship as Lived Research* (Bloomington: Indiana University Press), 35–53.

Hill Collins, P. (1997) 'Comments on Hekman's "Truth and Method": Where's the Power?', *Signs*, 22(21): 375–381.

Hundleby, C. (1997) 'Where Standpoint Stands Now', *Women & Politics*, 18(3): 25–43.

Intemann, K. (2000) 'Science and Values: Are Moral Judgments Always Irrelevant to the Justification of Scientific Claims?', *Philosophy of Science*, 68 (Proceedings) (3): S506–S518.

Kincaid, H. (2007) 'Contextualist Morals and Science', in H. Kincaid, J. Dupré, and A. Wylie (eds), *Value-Free Science? Ideals and Illusions* (Oxford: Oxford University Press), 218–236.

Kitcher, P. (1993) *The Advancement of Science* (Oxford: Oxford University Press).

Kourany, J. (2003) 'A Philosophy of Science for the Twenty-First Century', *Philosophy of Science*, 70: 1–14.

Kuhn, T. (1977) 'Objectivity, Value Judgment, and Theory Choice', in *The Essential Tension* (Chicago: University of Chicago Press).

Lacey, H. (1999) *Is Science Value-Free?* (London: Routledge).

Laudan, L. (1984) *Science and Values* (Berkeley: University of California Press).

Laudan, L. (2004) 'Cognitive, Social, and Epistemic Values', in P. Machamer and G. Wolters (eds), *Science, Values and Objectivity* (Pittsburgh: University of Pittsburgh Press).

Longino, H. (1990) *Science as Social Knowledge* (Princeton: Princeton University Press).

Longino, H. (1993) 'Subjects, Power, and Knowledge: Description and Prescription in Feminist Philosophies of Science', in L. Alcoff and E. Potter (eds), *Feminist Epistemologies* (New York: Routledge), 101–120.

Longino, H. (1995) 'Gender, Politics, and the Theoretical Virtues', *Synthese*, 104: 383–397.

Longino, H. (1996) 'Cognitive and Non-cognitive Values in Science: Rethinking the Dichotomy', in L. Hankinson Nelson and J. Nelson (eds), *Feminism, Science, and the Philosophy of Science* (Dordrecht: Kluwer Academic Publishers), 39–58.

Longino, H. (2002) *The Fate of Knowledge* (Princeton: Princeton University Press).

Miller, R. (1987) *Fact and Method: Explanation, Confirmation, and Reality in the Natural and the Social Sciences* (Princeton: Princeton University Press).

Mischel, W. (1958) 'Preference for Delayed Reinforcement: an Experimental Study of a Cultural Observation', *Journal of Abnormal and Social Psychology*, 56: 57–61.

Pinnick, C., N. Koertge, and R. Almeder (eds) (2003) *Scrutinizing Feminist Epistemology* (New Brunswick, NJ: Rutgers University Press).

Reichenbach, H. (1938) *Experience and Prediction* (Chicago: University of Chicago Press).

Rolin, K. (2006) 'The Bias Paradox in Feminist Standpoint Epistemology', *Episteme*, 1: 125–136.

Rose, H. (1987) 'Hand, Brain, and Heart: a Feminist Epistemology for the Natural Sciences', in S. Hardin and J. O'Barr (eds), *Sex and Scientific Inquiry* (Chicago: University of Chicago Press).

Schneider, S. H., K. Kuntz-Duriseti, and C. Azar (2000) 'Costing Non-linearities, Surprises and Irreversible Events', *Pacific and Asian Journal of Energy*, 10(1): 81–106.

Shiva, V. (2006) 'Women's Indigenous Knowledge and Biodiversity Conservation', in *Theorizing Feminisms* (Oxford: Oxford University Press), 238–243.

Smith, D. (1974) 'Women's Perspective as a Radical Critique of Sociology', *Sociological Inquiry*, 44: 7–14.

Smith, D. (1987) *The Everyday World as Problematic: a Feminist Sociology* (Boston: Northeastern University Press).

Smith, D. (1992) 'Sociology from Women's Experience: a Reaffirmation', *Sociological Theory*, 10(1): 88–98.

Smith, D. (1997) 'Comments on Hekman's "Truth and Method": Where's the Power?', *Signs*, 22(21): 392–398.

Solomon, M. (2001) *Social Empiricism* (Cambridge, MA: MIT Press)

Williams, M. (2001) *Problems of Knowledge: a Critical Introduction to Epistemology* (Oxford: Oxford University Press).

Wylie, A. (2001) 'Doing Social Science as a Feminist: the Engendering of Archaeology', in A. N. H. Creager, E. Lunbeck, and L. Schiebinger (eds), *Feminism in Twentieth Century Science, Technology, and Medicine* (Chicago: University of Chicago Press), 23–45.

Wylie, A. (2003) 'Why Standpoint Matters', in R. Figueroa and S. Harding (eds), *Science and Other Cultures: Issues in Philosophies of Science and Technology* (New York: Routledge), 26–48.

Wylie, A. and L. Hankinson Nelson (2007) 'Coming to Terms With the Values of Science: Insights From Feminist Science Studies Scholarship', in H. Kincaid, J. Dupré, and A. Wylie (eds), *Value-Free Science: Ideals and Illusions* (Oxford: Oxford University Press).

11

Naturalism and the Enlightenment Ideal: Rethinking a Central Debate in the Philosophy of Social Science

Daniel Steel[1]

1. Introduction

The naturalism versus interpretivism debate in the philosophy of social science is traditionally framed as the question of whether social science should attempt to emulate the methods of natural science. In this chapter I show that this manner of formulating the issue is problematic insofar as it presupposes an implausibly strong unity of method among the natural sciences. I propose instead that what is at stake in this debate is the feasibility and desirability of what I call the Enlightenment ideal of social science. I argue that this characterization of the issue is preferable, since it highlights the central disagreement between advocates of naturalism and interpretivism, makes connections with recent work on the topics of causal inference and social epistemology, while avoiding unfruitful comparisons between the social and natural sciences.

According to the traditional way of conceiving the naturalism versus interpretivism debate, *naturalists* are those who are in favour of social science emulating the methods of natural science, while *interpretivists* are those who argue against it. Thus, interpretivists maintain that unique features of human beings – such as the capacity for self-conceptualization and the ability to assign meaning to otherwise arbitrary objects or events – entail that the procedures of natural science are inapplicable to the social realm. Instead, interpretivists propose that social science should seek understanding, in a sense more akin to the interpretation of a literary text than explanation by reference to natural laws or causal mechanisms. In contrast, naturalists maintain that the social sciences can and should follow natural science standards. Moreover, naturalists question whether literary-style interpretive techniques can provide the understanding that interpretivists promise.

The traditional formulation of the naturalism versus interpretivism dispute implicitly presupposes a strong thesis concerning the unity of scientific method, according to which there is a set of uniform and specific rules that comprise the methods of all natural sciences. In this chapter, I argue that this presupposition is deeply problematic. Methodological rules general enough to apply to all natural sciences are typically too vague and minimal to engage with controversies concerning social science methods, while more precise and informative rules tailored to specific contexts differ from particle physics, to astronomy, to geology, to molecular biology, to evolutionary biology, and so on. If this is right, then little is to be gained by enunciating a list of putative natural science standards and asking whether the social sciences live up to them. For the rules of particular natural sciences are often inappropriate for social science, but by the same token the specific rules of one natural science will often be inappropriate for other natural sciences. On the other hand, basic standards of good evidence are often accepted by interpretivists as well as naturalists, and hence do not distinguish the positions typically defended under these banners and do little to resolve actual methodological disputes in social science.

In light of this, I propose that the naturalism versus interpretivism debate should be recast and reframed in a way that returns to what I regard as the historical origin of this issue. Leading Enlightenment figures such as Adam Smith, David Hume, and the Marquis de Condorcet regarded a science of society as the basis for rational social policy and the betterment of the human condition. Contemporary naturalists, though they often disagree with these figures on specific issues, nevertheless retain what I call the Enlightenment ideal of social science. Naturalists regard the discovery of causes of such social phenomena as economic growth, immigration, inequality, or crime as the primary goal of social science, and they think this knowledge is important precisely because of its relevance to urgent social problems. In contrast, one who rejects the Enlightenment ideal of social science could claim either that it is impossible to achieve or that it would not be desirable even if it were. For example, interpretivist social scientists such as Clifford Geertz often stress that knowledge of human society is inextricably local, while the Enlightenment ideal requires generalizable knowledge capable of grounding effective social policies. Thus, one central component of the naturalism debate in the philosophy of social science focuses on the topic of causal inference, a topic on which there is currently a vibrant literature in the philosophy of science. On the other hand, one might be suspicious of the Enlightenment ideal's conception of the 'betterment of the human condition'. Whose conception of 'betterment' is presumed here and how does that conception impact social science research? This challenge to the Enlightenment ideal raises some basic questions of social epistemology, which are also the focus of an active philosophy of science literature.

In what follows, I will elaborate and defend the claims made above and sketch some of their implications. I begin by arguing that acceptance of some basic standards of evidence is not what distinguishes naturalists from interpretivists, and consequently that this dispute must be understood in some other way. In the next section, I elaborate on the Enlightenment ideal and explain how it nicely highlights two central issues at play in discussions of naturalism in the philosophy of social science: causal inference and the role of social values in social science. The final two sections explain how these two topics are more adequately addressed given the formulation of the naturalism–interpretivism debate that I propose. In particular, formulating the issue in terms of the Enlightenment ideal of social science frees naturalism from the assumption that social science must discover laws of nature similar to those of physics as well as from a commitment to an outmoded distinction between the contexts of discovery and justification.

2. Natural science standards

An expression like 'natural science standards' might be taken to indicate at least two distinct types of thing. On the one hand, the standards in question might be relatively domain-specific, such as the double-blinded randomized controlled experiment as the gold standard in clinical research or the requirement that any acceptable theory of gravitation must postulate general laws that pass the three classical tests of General Relativity. On the other hand, the phrase might be taken to refer to such general rules as 'do not multiply entities beyond necessity' or 'severely test hypotheses'. If one takes very specific methodological rules as one's touchstone, it is easy to argue that social science fails to live up to natural science standards. For example, hypotheses in social science usually cannot be tested by means of randomized controlled experiments, and, unlike physics, social science does not appear to discover anything resembling universal laws of nature. Not surprisingly, defenders of naturalism tend to interpret 'natural science standards' as referring to rules of the latter, more abstract sort, while interpretivists often do the opposite.

The classic formulation of the naturalist position in social science is given by John Stuart Mill in his *System of Logic*. Mill writes that, 'The backward state of the Moral [i.e. human] sciences can only be remedied by applying to them the methods of Physical Science, duly extended and generalized' (1856: xii). Contemporary philosophers of social science state naturalism in a manner very similar to Mill (cf. Bhaskar, 2000; Kincaid, 1996; Papineau, 1978). For instance, Harold Kincaid, a leading contemporary proponent of naturalism, sums up the position as follows:

> (1) The social sciences can be good science by the standards of the natural sciences;
> (2) The social sciences can only be good science by meeting the standards of the natural sciences. (1996: 3)

Kincaid makes clear that he does not interpret natural science standards as specific methods in particular fields like oceanography or molecular genetics but rather as 'certain basic scientific virtues' that lie behind such methods (ibid.). Three virtues of scientific testing are discussed by Kincaid: *independent tests*, *fair tests*, and *cross tests* (1996: 51–53). Independent tests are ones in which the hypothesis being examined is not presupposed in the design of the test, and there is some independent reason to accept the assumptions upon which the test depends. Fair tests satisfy the further condition that the assumptions presumed in the test do not unfairly favor the hypothesis under examination over its competitors. Finally, cross testing a hypothesis means testing it in a variety of ways with several distinct types of data.

Kincaid's virtues of scientific testing are very sensible. In fact, it seems likely that many interpretivists would agree with them. After all, interpretivism is not synonymous with epistemological relativism. Interpretivism need not be conjoined with the view that assessing the merits of rival interpretations is a matter of ineffable and idiosyncratic intuition to which standards of evidence are irrelevant. For example, in response to the objection that conclusions drawn by interpretive anthropologists are often arbitrary (Jones, 1998), one anthropologist argues that such interpretations are constrained by considerations very similar to the sort enumerated by Kincaid (Segal, 1999). Similarly, in response to the suggestion that interpretations of social meanings are entirely subjective, Clifford Geertz writes, 'For a field of study which, however timidly (though I, myself, am not timid about the matter at all), asserts itself to be science, this just will not do' (1973: 24). Indeed, the concern to articulate objective standards of understanding is a central theme of Wilhelm Dilthey's essay, 'The Development of Hermeneutics',

Figure 11.1 Migrant labor and revolutionary nationalism

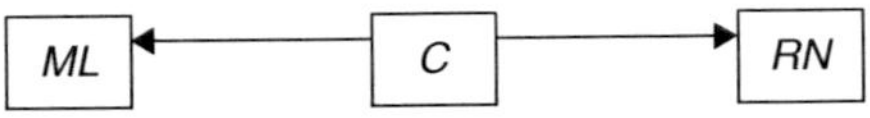

Figure 11.2 An alternative hypothesis

which is often regarded as the foundational document of interpretive social science.[2] Whatever naturalism is, it should be something that interpretivists would reject. Thus, formulating naturalism in terms of some very general standards of good evidence is problematic, since many interpretivists would accept such standards, too.

In addition, Kincaid's criteria of independent, fair, and cross testing are not sufficiently specific to engage with discussions of other methodological problems in social science, such as causal inference from observational data. Consider an example of social science research that Kincaid regards as exemplifying the virtues of independent, fair, and cross tests: Jeffery Paige's (1975) classic *Agrarian Revolution*. According to Kincaid:

> Paige's theory of agrarian movements certainly seems to have the traits of good science. It exhibits the *evidential* virtues summarized by independent, fair, and cross tests. (1996: 78; italics in original)

Paige proposes a theory about which types of agrarian social structures produce which types of conflict between cultivators and non-cultivators. Among the several hypotheses that comprise this theory, I consider just the one that can be represented in Figure 11.1. The variable *ML* indicates the presence or absence of an agrarian system in which landlords' income is based predominantly on land ownership (rather than commercial or industrial capital) and in which landlords rely on migrant labor. The variable *RN* indicates the presence or absence of revolutionary movements among migrant laborers that appeal to nationalism as a basis of solidarity. Thus, the hypothesis is that the presence of such an agrarian system is a cause of nationalist revolutionary movements. Paige presents two lines of statistical evidence for this hypothesis (1975: 104–108). First, there is a positive, statistically significant correlation between *ML* and *RN*.

Secondly, Paige argues that this correlation cannot be accounted for by the alternative hypothesis represented in Figure 11.2. In this graph, *C* is a variable indicating the presence or absence of a colonial regime. Paige's argument against this alternative is that a model in which *ML* and *C* influence *RN* interactively fits the data better than one in which they influence it additively (1975: 108). Moreover, in the interactive model, there is no statistically significant correlation between *C* alone and *RN*, although there is a significant correlation between *RN* and the interaction term for *C* and *ML* (ibid.: 109). Thus, Paige concludes that since *C* and *ML* interact to cause *RN*, the alternative represented in Figure 11.2 cannot be correct.

Although Paige's study may be a case of independent, fair, and cross testing, there are active debates in social science and philosophy about whether the sorts of methods he employs can in fact justify causal conclusions. Paige considers just two alternatives, and argues that the one in which *ML* is a cause of *RN* fits the data best. Yet this result does not eliminate other models that include further variables and in which *ML* is *not* a cause of *RN*. Suppose that Paige had pursued this possibility, and had drawn up a reasonably comprehensive list of potential common causes of *ML* and *RN*. Then he would have been confronted with the problem that increasing the number of variables rapidly generates a combinatorial explosion in the number of alternative models. It is a difficult question to what extent there are methods capable of reliably drawing informative causal conclusions in the face of such a vast array of alternatives.[3] The point here is definitely not to suggest that Paige's work is substandard: to the contrary, I agree with Kincaid that it is an example of careful and conscientious social science. The issue concerns the adequacy of methods available to social scientists for learning the causes and effects of social phenomena. And Kincaid's criteria of independent, fair, and cross tests simply do not engage with this issue.

The same point can be made with regard to methodological disputes concerning methods of causal inference used in econometrics. Such methods typically assume that one begins with some qualitative knowledge of the causal structure (say, on the basis of micro-economic theory). The inference problem then consists of providing quantitative estimates of causal impacts. A common objection to such methods is that they rely upon modeling assumptions whose accuracy is generally difficult to ascertain (cf. LaLonde, 1986). Assumptions are made about such things as the functional form of the relationships between causes and effect, the probability distribution of the error terms, and whether the error terms are independent. Since the accuracy of such assumptions is often quite uncertain, it is often difficult to assess which model one should choose. A response to such objections is that specification tests can be performed to assess whether the assumptions in the model are accurate (cf. Heckman and Hotz, 1989). An example of such a test monitors matched groups for some time prior to the implementation of the program to see if any significant differences in outcomes emerge. If not, the model passes the specification test. The utilization of specification tests presumably entails that the inference methods in question satisfy Kincaid's criteria of fair and independent testing. That is, the tests do not presuppose the accuracy of any particular hypothesis concerning quantitative causal effects, and the specification tests provide some independent basis for the assumptions upon which the test is premised. However, some object that such specification tests themselves depend upon questionable assumptions[4] and are not always effective in weeding out models that generate inaccurate estimates (Friedlander and Robins, 1995). Moreover, some studies find that sophisticated statistical methods intended to produce balanced comparison groups are often ineffective at reducing bias, sometimes significantly increase it (Michalopoulos et al., 2004), and do far less to reduce bias than simply selecting the comparison group from an adjacent locality (Friedlander and Robins, 1995; Michalopoulos et al., 2004).

In sum, many interpretivists would accept Kincaid's criteria of independent, fair, and cross testing, and those standards do little to address other methodological disputes in social science. Hence, showing that social science research is sometimes in accord with these criteria is not a strong argument for naturalism. On the other hand, attempting to refute naturalism on the grounds that it fails to live up to very specific standards of natural science faces the objection that these more specific rules do not hold for all natural sciences either. For example, critiques of naturalism often maintain that certain features of human beings make it impossible for there to be universal laws of society.[5] Yet a defender of naturalism can respond to this by pointing out that some natural sciences (e.g. molecular biology) do not discover, nor even seem to attempt to discover, universal laws of nature.[6] For example, several authors have argued that explanations in many biological sciences are based primarily on mechanisms rather than laws (cf. Craver, 2007; Glennan, 2005; Machamer et al., 2000). Similarly, controlled experiments are often not a feasible means of testing hypotheses in social science, but the same can be said for astronomy, geology, and evolutionary biology. The conclusion to be drawn from this is that little is gained from drawing up a list of supposed standards of natural science and asking whether social science can abide by them. Consequently, if there is an issue in play in the naturalism versus interpretivism dispute that is worth discussing, it must be formulated in other terms. In the next section, I develop a proposal about what those terms might be.

3. The Enlightenment ideal of social science

Daniel Little also argues that the diversity of methods in natural and social science undermines a basic presupposition of the naturalism versus interpretivism debate (1991: ch. 11). However, the conclusion he draws from this is that naturalism and interpretivism are both somewhat wrong and somewhat right: sometimes social science utilizes methods with clear parallels to those found in natural science, other times it uses very distinctive methods; sometimes social science aims to discover causal relationships, other times it aims solely for interpretive understanding.[7] Although I am sympathetic to this ecumenical perspective, I also think that it fails to address a fundamental disagreement that underlies the naturalism versus interpretivism debate.

I claim that naturalism is ultimately motivated by what I call the *Enlightenment ideal* that a science of society would, by discovering causes and effects of social phenomena, serve as a basis for rational policies that address important social problems such as unemployment, poverty, and crime (cf. Kincaid, 1996: xv).[8] As understood here, this ideal is not linked to particular social or political theories of specific Enlightenment figures such as Adam Smith or the Marquis de Condorcet.[9] For example, the Enlightenment ideal of social science does not include a commitment to laissez-faire economic policies, which many contemporary naturalists would reject. The Enlightenment ideal as understood here is also not committed to a belief in some utopian future state, which according to some historians was a guiding light of many leading Enlightenment thinkers (cf. Becker, 1932). Contemporary naturalists are more likely to view social science as making a positive contribution in a more piecemeal fashion of continuous tinkering in a social world inevitably beset by numerous challenges and problems.[10] What contemporary naturalists do share with Enlightenment social theorists is the conviction that social science can deliver generalizable knowledge that can serve as a basis for improving society.

It is easy to see how the Enlightenment ideal, if combined with the implicit presumption of a strong thesis concerning the unity of scientific method, would lead directly to the formulation of naturalism described in the previous section. Given a strong thesis concerning the unity of scientific method, a straightforward way to pursue the Enlightenment ideal of social science would be to identify the scientific discipline that has achieved the greatest success and attempt to emulate its methods. For instance, if physics is the most successful science, then the unity of scientific method thesis would entail that social science should pursue a method modeled on physics. Once the discussion is framed in this manner, a failure of social science to fit the model of scientific method exemplified by physics could be taken as evidence that social science cannot be genuine science. In response, those who wish to defend naturalism could insist that scientific method in fact consists only of some general standards of evidence, such as 'severely test hypotheses'. And that leaves you exactly where the previous section began. Yet, one who is skeptical of strong theses concerning the unity of scientific method would regard the question of whether social science can emulate some model of scientific method based on physics as misguided and beside the point. In the absence of such assumptions, the question of whether the Enlightenment ideal of social science is attainable is independent of whether social science resembles one or another natural science.

Let us consider, then, what the substantive issues in the naturalism versus interpretivism debate would be if it is framed as I propose here. Plainly, the feasibility of the Enlightenment ideal of social science depends on the existence of reliable methods for discovering causes and effects of social phenomena. Difficult issues of social policy almost invariably turn on hard to answer questions of cause and effect. For example, do federal budget deficits inhibit economic growth? Do gun control laws reduce violent crime? What effects would a particular welfare reform policy have upon poverty? What effect does trade liberalization have on global poverty? Social science can achieve the Enlightenment ideal only if it is capable of providing accurate answers to such questions. But the Enlightenment ideal also contains a strong normative element: the knowledge it seeks is important because of its potential to provide a basis for policies that would lessen social ills. That raises questions about the conception of social improvement associated with naturalism and its connection to social science research. Defenses of naturalism typically devote less attention to the normative side of the Enlightenment ideal, perhaps because they implicitly regard such issues as separate from questions of scientific method and evidence. For example, according to one well-known argument, political and social values pertain only to the context of discovery and not to the context of justification.[11] However, a number of philosophers have cast doubt on the distinction between the contexts of discovery and justification as a means for sequestering epistemology from social and political values (cf. Kitcher, 2001; Longino, 2002; Okruhlik, 1994). Thus, I think that a defense of naturalism in the philosophy of social science simply cannot sidestep the question of the connection between values and method. Moreover, these two issues – causal inference and the connection between values and methods – correspond directly to two central lines of criticism of naturalist approaches to social science. In the following two sections, I consider these issues in turn.

4. Interpretation and causation

I argued above that acceptance of basic standards of good evidence is not a feature that distinguishes interpretivists from naturalists. An advantage of formulating the dispute in terms of the Enlightenment ideal, then, is that it highlights issues about which naturalists and interpretivists actually do diverge. One such issue concerns the ability of social science to discover generalizable causal relationships. Traditional defenses of naturalism tend to frame this issue as the question of whether there are laws of social science that resemble laws in natural sciences such as physics. In this section, I argue that this traditional manner of defending naturalism rests on an implicit assumption of a strong thesis of the unity of scientific method and tends to obscure the pertinent issues concerning causal inference and generalization. I explain how construing naturalism in terms of the Enlightenment ideal of social science results in a better formulation of this issue.

A perennial topic in Anglo-American philosophy of social science in the latter half of the twentieth century was whether explanations of human action are inherently non-causal (Davidson, 1980; Henderson, 1993; Taylor, 1971; Winch, 1958). Clearly, the Enlightenment ideal of social science would be for naught if this were so. However, one might regard the Enlightenment ideal as unattainable even if human actions and institutions have causes. For the Enlightenment ideal requires causal knowledge of social phenomena that is capable of guiding social policy, and such knowledge must be generalizable. For example, an assessment of the effects of a change in social welfare policies must generalize from some locations to others, from smaller to larger scales, and from past to future. One who rejects the Enlightenment ideal of social science, then, might agree that social phenomena have causes that can sometimes be discovered but claim that such causal knowledge is inevitably local and not generalizable in the way that the naturalist requires. In fact, the emphasis on the locality of social knowledge is one of the central themes in the writings of a leading interpretivist figure of the latter twentieth century, Clifford Geertz.

Reflecting on the interpretive turn in social science that he helped launch, Geertz refers to 'the growing recognition that the established approach to treating [cultural] phenomena, laws-and-causes social physics, was not producing the triumphs of prediction, control, and testability that had for so long been promised in its name' (1983: 3). Clearly, this is a statement of skepticism of naturalism understood as social science modeled after physical science. But unlike some interpretivists, such as Peter Winch, Geertz does not claim that it is a conceptual confusion to speak about causes and effects of social phenomena. For example, Geertz writes:

> Social events do have causes and social institutions effects; but it just may be that the road to discovering what we assert in asserting this lies less through postulating forces and measuring them than through noting expressions and inspecting them. (1983: 34)

Moreover, I think that a careful reading of Geertz's actual work shows a genuine concern with questions of cause and effect. For example, consider his well-known essay, 'Deep Play: Notes on the Balinese Cockfight' (1973: ch. 15).

As Michael Martin (1993: 273–274) notes, commentators on Geertz's essay have produced rather different accounts of his interpretation of the Balinese cockfight. Some read Geertz as asserting that the Balinese cockfight performs the therapeutic function of allowing for ritual expressions of latent conflicts (Rabinow and Sullivan, 1987). Others interpret Geertz as saying that the Balinese cockfight represents the struggle between good and the 'Powers of Darkness' (Little, 1991). One commentator interprets Geertz as asserting that the Balinese cockfight symbolizes class divisions within Balinese villages (Roseberry, 1982). Others interpret Geertz in a way that combines the latter two readings (cf. Jones, 1998). Although most of these readings identify genuine themes in Geertz's essay, I think that they all miss his central thesis, which concerns the notion of 'deep play' mentioned so prominently in the title of his article. What Geertz proposes is a pair of causal hypothesis. The first concerns factors that make a cockfight deep rather than shallow and uninteresting, while the second is a claim about the effects of deep fights upon Balinese observers and participants.

The 'deepness' of a match has to do with the level of excitement it generates, which is normally expressed by the quantity of money wagered on it. According to Geertz, a typical fight would involve two sorts of bet: the center bet and side bets (1973: 425–426). The center bet is contracted between the two men whose roosters are to engage in mortal combat, and this bet is always made at even money odds. Side bets are made among the spectators and are contracted at odds ranging from ten-to-nine to two-to-one. Geertz then makes the following generalization: 'the higher [the stakes of] the center bet, the more likely the match will in actual fact be an equal one' (1973: 429). Geertz, who recorded the stakes of center bets and outcomes for fifty-seven fights, provides statistical data to support this claim (1973: 430). Geertz also reports that the higher the stakes of the center bet, the greater the volume of side betting and the more even the odds of the side bets.

High stakes matches involving evenly paired cocks and generating considerable side betting are what Geertz means by 'deep play'. According to Geertz, deep matches are what Balinese cockfighting connoisseurs regard as the truly interesting and important kind of fight. Thus, Geertz wishes to understand what makes some matches deep and others not so. Geertz suggests that the greater stakes involved are only an ancillary part of the explanation:

> The center bet is a means, a device for creating 'interesting', 'deep' matches, *not* the reason, or at least not the main reason, *why* they are interesting, the source of their fascination, the substance of their depth. The question of why such matches are interesting – indeed, for the Balinese, exquisitely absorbing – takes us out of the realm of formal concerns into more broadly sociological and social-psychological ones, and to a less purely economic idea of what 'depth' in gaming amounts to. (1973: 432; italics in original)

According to Geertz, Balinese villages are typically divided into several 'large, patrilineal, partly endogamous descent groups which are constantly vying with one another and form the major factions of the village' (1973: 437). Moreover, the center stakes in deep matches are normally too great to be raised by a single person, and consequently the wager on each side is pooled among a group of people. Those placing money on one side would be primarily composed of members of one faction, while those supporting the other side would be drawn from an opposing faction (1973: 437). Geertz's hypothesis, then, is that the enmity and social status of the two factions are the primary variables explaining the deepness of the match. Geertz states this hypothesis quite clearly by way of the following schema:

THE MORE A MATCH IS . . .

1. Between near status equals (and/or personal enemies)
2. Between high status individuals

THE DEEPER THE MATCH. (1973: 441)

The deepness of the match is then associated with the factors described above: evenly matched cocks, higher stakes, greater interest, etc. According to Geertz, conditions 1 and 2 are what make a fight deep and hence interesting to the Balinese; when these conditions are lacking, the fight will stimulate less interest, will have lower stakes, and so on than otherwise. The above conclusion is preceded by a list of seventeen facts given as supporting evidence (1973: 437–440). Geertz's second hypothesis concerns the psychological effects upon Balinese of attending and participating in cockfights (1973: 449). For example, Balinese might learn to identify with their social faction and to react with intensely joyful emotions when their faction bests one of its rivals. I will not have much to say about this second hypothesis, since the first is sufficient for the points I wish to make in this section.

The above example shows that interpretivists sometimes propose causal hypotheses and feel compelled to provide evidence for them. Indeed, Geertz's evidence for his hypothesis might be reasonably argued to satisfy Kincaid's criteria of good testing. Thus, what distinguishes naturalists from an interpretivist like Geertz is neither an interest in causes of social phenomena nor the acceptance of some basic standards of evidence. The difference, I think, has to do with the question of the extent to which causes can be generalized from one social context to another. For example, Geertz writes that the aim of social theory as he conceives it 'is not to codify abstract regularities but to make thick description possible, not to generalize across cases but within them' (1973: 26). Geertz goes on to develop his notion of generalization 'within cases' by means of an analogy with medical diagnosis. I will not explore difficulties confronting Geertz's positive proposal here.[12] Instead, I wish to consider the link between naturalism and the question of generalizing causal claims from one social context to another.

One way to answer the objection that causal generalizations in social science are inextricably tied to specific social contexts would be to point to examples of general and informative causal claims in social science for which there is considerable evidence. For example, Kincaid makes precisely this sort of argument with regard to the law of supply and demand in economics (1996: 232–247). However, such defenses of naturalism also typically presume that the general causal claims must qualify as laws. Since it is doubtful whether there any laws of social science that are true everywhere and always without exception, it is often suggested that social science laws are 'ceteris paribus' laws (cf. Hausman, 1992; Kincaid, 1996: ch. 3). The phrase 'ceteris paribus' is Latin for 'other things being equal', and a ceteris paribus law (hereafter, cp law) is a generalization with a potentially open ended qualifying phrase attached to it that would indicate circumstances in which the law might not hold. Defenses of cp laws in social science, then, constitute an attempt to show that the social sciences have laws that are analogous to physical laws. Thus, defenders of cp laws often argue that fundamental laws of physics are cp laws, too (cf. Kincaid, 1996: 32).

I think that there are a number of shortcomings with appealing to cp laws to argue that causal knowledge can be generalized across social contexts. One problem is that it raises a variety of questions that simply are not germane to the issue, such as the question of whether laws are ceteris paribus 'all the way down' to physics. That is a very contentious issue (cf. Earman and Roberts, 1999) about which philosophers of social science are not likely to have much useful to say. But most importantly, it is irrelevant to the challenge to the Enlightenment ideal posed by interpretive social scientists such as Geertz. What matters is the extent to which causal knowledge can be generalized in social science, not whether generalizations in social science have the same form as those in physics. A fixation with cp laws is, I think, a holdover of implicitly assuming a strong thesis of the unity of scientific method while regarding physics as the paradigm science.

A second problem with the cp law approach is that it tends to obscure the diversity of generalizations found in social science, as well as in other sciences such as biology. The effort to assimilate these various types of generalization into a model inspired by examples from physics results in unnecessary confusions. For example, Gerhard Schurz (2002) shows that at least four significantly distinct types of claims are often referred to by the expression 'cp law'. Perhaps the most commonly encountered interpretation of cp laws in the philosophical literature is what Schurz characterizes as *indefinite exclusive* (cf. Fodor, 1991; Hausman, 1992; Pietroski and Rey, 1995).[13] Indefinite exclusive cp laws assert that a generalization holds without exception under an open-ended set of circumstances that cannot be fully specified. Of the four types of cp laws described by Schurz, the indefinite exclusives are by far the most problematic. For example, Schurz shows that indefinite exclusive cp laws are 'almost empty' in the sense of being equivalent to claims about determinism. For example, 'Ceteris paribus, all Fs are Gs', if interpreted as an indefinite exclusive cp law, is equivalent to saying that for everything that is F, there are deterministic causes of whether it is G. Consequently, it is difficult to see how indefinite exclusive cp laws could be tested by empirical means. Defenders of indefinite exclusive cp laws sometimes respond to such objections by claiming that they can be tested by randomized controlled experiments (Hausman, 1992: 137; Kincaid, 1996: 67–68). However, I think this claim is a mistake that results from confusing indefinite exclusive with what Schurz terms *comparative* cp laws. In contrast to the exclusives, a comparative cp law does not require that interfering factors are absent but instead that factors relevant to the outcome are distributed uniformly between groups that differ with respect to the cause. Comparative cp laws correspond to what Judea Pearl (2000: 70) calls a causal effect, which specifies the probability distribution of one variable *Y* conditional on another *X* whose values are set by an ideal intervention. Causal effects can be estimated in randomized experiments as well as from observational data in some circumstances (cf. Scheines, 2005). But given that indefinite exclusive cp laws are equivalent to claims about determinism, it is clear that they cannot be tested by randomized controlled experiments. For whatever merits randomized experiments might have, they certainly cannot answer metaphysical questions about determinism.

Finally, the emphasis on cp laws obscures the fact that laws on the model of physical science may not be the only possible basis for generalizing causal claims across contexts. For example, a popular idea in the current philosophy of biology literature is that explanations in biology are much more frequently based on mechanisms than laws (cf. Machamer et al., 2000). Moreover, knowledge of mechanisms, of factors that may interfere with them, and of how they are likely to differ or be similar in various contexts can be useful for extrapolating claims about causal effects, for instance, from animal models to humans (Steel, 2008). Since mechanisms have also attracted some interest within social science (cf. Hedström and Swedberg, 1998), this naturally raises the question of whether a mechanisms-based approach to extrapolation might work in social science, too (cf. Guala, 2005; Steel, 2008: ch. 8). Showing that this is indeed possible would be one way to answer the charge that knowledge of causes of social phenomena cannot be generalized from one context to another. And it is irrelevant to that project whether mechanisms and causal effects can be lumped into some category of 'cp law' along with fundamental laws of physics.

In sum, framing the naturalism debate in terms of the Enlightenment ideal highlights one of the central issues that divides interpretivists from naturalists, namely, the extent to which general causal knowledge is attainable in social science. Although defenses of naturalism formulated in the traditional manner do address this issue, they do so in a way that focuses attention on tangential questions that reflect an implicit assumption of a strong thesis of the unity of scientific method and a regard for physics as the paradigm science. In particular, these discussions assume that general causal claims in social science must resemble those in physics, an assumption that leads directly to the emphasis on cp laws. But if no such strong thesis of the unity of scientific method is presupposed, then a focus on cp laws is neither necessary nor even helpful. It is not necessary, since there is no reason to suppose that the methods best suited for social science must be substantially the same as those in physics. And it is unhelpful in that the emphasis on cp laws tends to conceal the diverse array of methods and causal claims available to social scientists. In short, it matters crucially to the Enlightenment ideal that it is possible for social science to acquire general causal knowledge, but not whether social science does this in a way that fits some supposed ideal model of physical science.

5. The betterment of society

There is a long-running dispute in philosophy of science as to whether scientific research can be insulated from social, moral, and political values and indeed whether it is even a good idea to try to do so (cf. Harding, 1986, 1998; Kitcher, 2001; Longino, 1990, 2002; Nagel, 1961). The classic defense of naturalism argues that non-epistemic values can, at least in principle, be excluded from social science research and that every effort should be made to keep them apart in practice (Kincaid, 1996: 45–46; Nagel, 1961: ch. 13; Papineau, 1978). Arguments for this position typically appeal to some form of the distinction between the contexts of discovery and justification. According to this line of reasoning, while non-epistemic values may come into play in the process of choosing research topics and constructing hypotheses or theories (the context of discovery), they are not relevant to questions about whether or not those hypotheses or theories are supported by the evidence (the context of justification). However, the context-of-discovery-versus-justification distinction has come under a considerable degree of criticism in the philosophy of science literature within the past twenty to thirty years. In this section, I suggest that one significant advantage of construing naturalism in terms of the Enlightenment ideal is that it frees naturalism from the context-of-discovery-versus-justification dichotomy. I also sketch some ways the issue of non-epistemic values in social science research can be more fruitfully addressed from a naturalist perspective.

While the traditional defense of naturalism seeks to isolate social science from all social values, my formulation of the issue explicitly emphasizes the link between social science and a commitment to social improvement. At the heart of the Enlightenment ideal is the thesis that social science should discover knowledge that will contribute to the betterment of society. Given this perspective, it would be strange indeed to insist that social science must be strictly isolated from social values, since without some connection of that kind social science would be an enterprise with no purpose. Thus, if naturalism is understood in terms of the Enlightenment ideal, there should be no temptation to sweep social values under the rug with a dubious context-of-discovery-versus-justification broom. Instead, the connection between social science and social values would have to be confronted squarely.

One difficulty with the context-of-discovery-versus-justification distinction is that there are now several works explicitly aimed at providing an account of the logic of scientific discovery (cf. Kelly, 1996: ch. 9; Langley et al., 1987), as well as case studies examining the logic of the discovery process in specific scientific domains (Darden and Craver, 2002). Thus, it is doubtful that discovery, in contrast to justification, is some ineffably intuitive process. On the contrary, the two seem symmetrical: both are amenable to logical analysis and in practice both can be influenced by social and political values. But perhaps the most significant problem with the context-of-discovery-versus-justification distinction for present purposes is that it would not be sufficient to sequester scientific results from social values even if it were granted. As Kathleen Okruhlik (1994) points out, substantial biases in scientific results could remain even if the influence of social and political values could be restricted to the creation of hypotheses. In particular, Okruhlik is concerned with the situation in which all hypotheses considered in a particular field share a common presupposition that reflects a pervasive gender bias, for instance, that women are naturally passive and nurturing. If only hypotheses that presuppose this are considered, then that gender bias will be reflected in the end result even if choices among these hypotheses are made in a way that is entirely unaffected by any social values.

The issue of political or moral values in social science research receives more attention in critiques of positivism than in the literature on interpretivism (cf. Steinmetz, 2005). Although 'positivism' is rarely clearly defined in this literature, it is usually taken to include some commitment to the exclusion of non-epistemic values in the assessment of social theories. For example, this is the central issue in Sandra Harding's defense of standpoint epistemology in social science (Harding, 2003, 2005). Standpoint epistemology rests on the premise that members of oppressed minority groups may have distinctive insights into the workings of social systems. In Harding's words, 'Standpoint projects use the distinctive ways a group is oppressed, exploited, or dominated to reveal the "mechanisms" by which elite groups maintain their dominance' (Harding, 2005: 355). There a number of objections to standpoint epistemology that I will not rehearse here (cf. Polhaus, 2002; Rolin, 2006). For present purposes, the primary difficulty with standpoint epistemology is that it provides very little guidance on the question of what the role of political values in social science research should be. Granted, a person who occupies one social position may be apt to have insights that would be unlikely to occur to a person from a different background. But we surely would not want to conclude from this that, for instance, whenever a white male economist disagrees with a Chinese-American female economist, the Chinese-American female economist must be correct. Of course, I do not mean to suggest that this is the sort of thing that Harding is proposing. Instead, the point is that Harding's standpoint epistemology needs to be augmented with some account of how insights stemming from differing political perspectives can be usefully integrated in social science.

A central component of a naturalistic approach to this issue is that non-epistemic values should not be allowed to influence social science research in a way that undermines basic standards of evidence, such as those advocated by Kincaid. Such a commitment would not typically be shared, I think, by post-modernist critics of positivism. Traditionally, naturalists took upholding standards of scientific evidence to mean that non-epistemic values should, at least in principle, be excluded entirely. However, I think that a survey of the contemporary literature on social epistemology should make clear that this is not the only possible option (cf. Goldman, 1999; Kitcher, 2001; Longino, 1990, 2002; Solomon, 2001). For example, several authors have suggested that scientific communities are more likely to produce reliable knowledge if they are organized in a way that promotes the free exchange of ideas among socially diverse members who are treated as intellectual equals. Although its roots trace back to John Stuart Mill and it was revived in the twentieth century by Karl Popper, Helen Longino is perhaps the best-known contemporary exponent of the idea that more open and equal social organization promotes better science. Central to Longino's proposal is an outline of the type of social organization that is capable of generating 'transformative criticism', which she regards as essential for scientific objectivity. Longino writes:

> Scientific communities will be objective to the degree that they satisfy four criteria necessary for achieving the transformative dimension of critical discourse: (1) there must be recognized avenues for the criticism of evidence, of methods, and of assumptions and reasoning; (2) there must exist shared standards that critics can invoke; (3) the community as a whole must be responsive to such criticism; (4) intellectual authority must be shared equally among qualified practitioners. (1990: 76)

Longino's thought is that epistemic communities that satisfy the above four requirements are likely to generate reliable knowledge, provided that a diverse array of viewpoints are present to sustain a vigorous discourse. I think that a naturalist should regard Longino's proposal as an appealing point of departure for addressing concerns about epistemological standpoints. However, since an explicit rejection of the context-of-discovery-versus-justification distinction is a fundamental aspect of Longino's account, adopting her approach would require abandoning the traditional naturalist position on the role of non-epistemic values in social science.

One advantage of this approach is that it would fruitfully engage naturalism with the current literature on social epistemology. The four criteria Longino articulates are quite appealing, and thus conflicts among those criteria or between them and other seemingly reasonable criteria for the effective organization of epistemic communities highlight important issues for social epistemology. For example, the emphasis on the value of a diverse range of viewpoints is very likely to conflict with Longino's criterion (2), which requires shared standards of criticism. Standards for evaluating arguments and reasoning are one of the things that people from socially and intellectually diverse backgrounds are likely to disagree about. Yet Longino surely is correct that some common ground is necessary for a useful discussion *and* that diversity of opinion is an essential ingredient in the advancement of knowledge. That leaves a crucial open question for social epistemology: how should these two conflicting criteria – the need for diversity and common ground – be balanced (cf. Solomon, 2006)? There are also conflicts between Longino's four criteria and other reasonable ideas about the sorts of social organization that promote knowledge. One appealing idea is that the scientific community ought to reward scientists whose work advances existing knowledge. Appointments to positions at prestigious universities and professional associations, such as the National Academy of Sciences, should be preferentially granted to scientists who develop new ideas, experiments, or techniques that solve previously unresolved problems within their fields (Strevens, 2003). Yet any such system of rewarding innovative work will inevitably lead to inequalities of intellectual authority, since the opinions of those who have achieved greater prestige will tend to carry more weight. Again, this conflict between appealing criteria for effective organization of an epistemic community raises another open challenge for social epistemology. Longino's account also does not address the question of how funding for scientific research should be distributed so as to best promote valuable social aims. That issue is important both for social epistemology and for the Enlightenment ideal of social science. Philip Kitcher (2001) addresses the issue of distribution of research funding as part of what he terms 'well-ordered science', a proposal which I also regard as a good point of departure for further developments of the connection between social values and social science research.

Traditional defenses of naturalism typically invoke a distinction between the contexts of discovery and justification to argue that social science research can remain unsullied by social and political values. In contrast, naturalism construed in terms of the Enlightenment ideal of social science would directly engage with difficult issues arising from the interaction of social science and social values. This approach is preferable because it does not rely on a dubious context-of-discovery-versus-justification dichotomy and because it engages with productive discussions currently underway in the literature on social epistemology.

6. Conclusion

The dispute concerning naturalism in the philosophy of social science has traditionally been posed in terms of the question of whether social science can and must follow the standards of the natural sciences. I have argued that this formulation of the issue is problematic in that it presumes an overly strong thesis about the unity of scientific method. Social science often fails to satisfy specific rules and standards of particular natural sciences, but the same is also true among distinct natural sciences themselves. And basic standards of good evidence that apply to all natural sciences are likely to be accepted by critics of naturalism. In light of this, I propose that naturalism be understood in reference to the Enlightenment ideal of a science of society that would discover causes and effects of social phenomena capable of supporting rational social policies. This formulation of the issue is preferable for several reasons. It highlights the issues about which naturalists and their critics actually disagree, and it frames those issues in a manner that avoids digressions on such topics as whether social science has laws that resemble those of physics. Moreover, it links disputes about naturalism to lively controversies elsewhere in the philosophy of science literature, such as social epistemology. Finally, I think that framing naturalism in terms of the Enlightenment ideal redirects the debate back to the important issues that motivated it in the first place.

Notes

1. I would like to thank Harold Kincaid for serving as commentator on an ancestral draft of this chapter at the 2005 Pacific APA. I would also like to thank Francesco Guala, P. D. Magnus, and an anonymous reviewer for helpful suggestions and commentary.
2. See Dilthey (1976).
3. See Glymour et al. (1987: 6–7) for a forceful statement of this challenge and its relevance to the reliability of studies like Paige's. Causal inference algorithms found in Neopolitan (2004), Pearl (2000), and Spirtes et al. (2000), are intended to provide means for searching among very large sets of alternative causal hypotheses.
4. For instance, the specification test mentioned above assumes that groups evenly matched at the earlier time will continue to be evenly matched at the later time (cf. Heckman and Hotz, 1989: 666).
5. For example, see Searle (1984: ch. 5) and Taylor (1971).
6. For instance, Kincaid (2000: S673) makes this argument.
7. Roth (1987) defends a similarly pluralistic conception of social science.
8. Bhaskar (2000) also associates naturalism with the Enlightenment.
9. See Rothschild (2001) and Schabas (2005) for excellent discussions of the economic and social theories of these and other Enlightenment figures.
10. For example, Popper explicitly associates this 'piecemeal tinkering' approach to social reform with the naturalist approach to social science (2002: ch. III).
11. See Ernest Nagel (1961: 485–502) for a classic and Kincaid (1996: 43–47) for a more recent defense of this position with respect to social science.
12. For example, Geertz's analogy with medical diagnosis seems rather problematic, since such diagnoses are based on knowledge about the causes and effects of various diseases that does generalize across cases.

13. The three other types of cp laws distinguished by Schurz are normic, definite exclusive, and comparative. For versions of the normic interpretation, see Lange (2002) and Schurz (2002), while for a version of the definite exclusive interpretation see Cartwright (1983: 45).

References

Becker, C. (1932) *The Heavenly City of the Eighteenth-Century Philosophers* (New Haven, CT: Yale University Press).

Bhaskar, R. (2000 [1979]) *The Possibility of Naturalism* (New York: Routledge).

Cartwright, N. (1983) *How the Laws of Physics Lie* (Oxford: Oxford University Press).

Craver, C. (2007) *Explaining the Brain* (New York: Oxford University Press).

Darden, L. and C. Carl (2002) 'Strategies in the Interfiled Discovery of the Mechanism of Protein Synthesis', *Studies in History and Philosophy of Biological and Biomedical Science*, 33: 1–28.

Davidson, D. (1980) *Essays on Actions and Events* (Oxford: Oxford University Press).

Dilthey, W. (1976) *Selected Writings*, ed. and trans. H. P. Rickman (Cambridge: Cambridge University Press).

Earman, J. and J. Roberts (1999) '*Ceteris Paribus*, There are no Provisos', *Synthese*, 118: 439–478.

Fodor, J. (1991) 'You Can Fool Some of the People All of the Time, Everything Else Being Equal: Hedged Laws and Psychological Explanations', *Mind*, 100: 19–34.

Friedlander, D. and P. Robins (1995) 'Evaluating Program Evaluations: New Evidence on Commonly Used Nonexperimental Methods', *American Economic Review*, 85: 923–937.

Geertz, C. (1973) *The Interpretation of Cultures* (New York: Basic Books).

Geertz, C. (1983) *Local Knowledge: Further Essays in Interpretive Anthropology* (New York: Basic Books).

Glennan, S. (2005) 'Modeling Mechanisms', *Studies in History and Philosophy of Biological and Biomedical Science*, 36: 443–464.

Glymour, C., R. Scheines, P. Spirtes, and K. Kelly (1987) *Discovering Causal Structure: Artificial Intelligence, Philosophy of Science, and Statistical Modeling* (Orlando, FL: Academic Press).

Goldman, A. (1999) *Knowledge in a Social World* (Oxford: Oxford University Press).

Guala, F. (2005) *The Methodology of Experimental Economics* (Cambridge: Cambridge University Press).

Harding, S. (1986) *The Science Question in Feminism* (Ithaca: Cornell University Press).

Harding, S. (1998) *Is Science Multi-Cultural?* Bloomington: Indiana University Press).

Harding, S. (2003) 'How Standpoint Methodology Informs Philosophy of Social Science', in S. Turner and P. Roth (eds), *The Blackwell Guide to the Philosophy of Social Sciences* (Oxford: Blackwell), 291–310.

Harding, S. (2005) 'Negotiating with the Positivist Legacy: New Social Justice Movements and a Standpoint Politics of Method', in G. Steinmetz (ed.), *The Politics of Method in the Human Sciences: Positivism and its Epistemological Others* (Durham, NC: Duke University Press), 346–365.

Hausman, D. (1992) *The Inexact and Separate Science of Economics* (Cambridge: Cambridge University Press).
Heckman, J. and J. Hotz (1989) 'Choosing among Alternative Nonexperimental Methods for Estimating the Impact of Social Programs: the Case of Manpower Training', *Journal of the American Statistical Association*, 84: 862–874.
Hedström, P. and R. Swedberg (eds) (1998) *Social Mechanisms: an Analytical Approach to Social Theory* (Cambridge: Cambridge University Press).
Henderson, D. (1993) *Interpretation and Explanation in the Social Sciences* (Albany, NY: SUNY Press).
Jones, T. (1998) 'Interpretive Social Science and the "Native's Point of View": a Closer Look', *Philosophy of the Social Sciences*, 28: 32–68.
Kelly, K. (1996) *The Logic of Reliable Inquiry* (Oxford: Oxford University Press).
Kincaid, H. (1990) 'Defending Laws in the Social Sciences', *Philosophy of the Social Sciences*, 20: 56–83.
Kincaid, H. (1996) *Philosophical Foundations of the Social Sciences: Analyzing Controversies in Social Research* (Cambridge: Cambridge University Press).
Kincaid, H. (2000) 'Global Arguments and Local Realism about the Social Sciences', *Philosophy of Science*, 67 (Proceedings): S667–S678.
Kitcher, P. (2001) *Science, Truth and Democracy* (New York: Oxford University Press).
LaLonde, R. (1986) 'Evaluating the Econometric Evaluations of Training Programs with Experimental Data', *American Economic Review*, 76: 604–620.
Lange, M. (2002) 'Who's Afraid of *Ceteris-Paribus* Laws? Or: How I Stopped Worrying and Learned to Love Them', *Erkenntnis*, 57: 131–147.
Langley, P., H. Simon, G. Bradshaw, and J. Zytkow (1987) *Scientific Discovery: Computational Explorations of the Creative Process* (Cambridge, MA: MIT Press).
Little, D. (1991) *Varieties of Social Explanation: an Introduction to the Philosophy of Social Science* (Boulder, CO: Westview Press).
Longino, H. (1990) *Science as Social Knowledge: Values and Objectivity in Scientific Inquiry* (Princeton: Princeton University Press).
Longino, H. (2002) *The Fate of Knowledge* (Princeton: Princeton University Press).
Machamer, P., L. Darden, and C. Craver (2000) 'Thinking about Mechanisms', *Philosophy of Science*, 67: 1–25.
Martin, M. (1993) 'Geertz and the Interpretive Approach in Anthropology', *Synthese*, 97: 269–286.
Michalopoulos, C., H. Bloom, and C. Hill (2004) 'Can Propensity-Score Methods Match the Findings from a Random Assignment Evaluation of Mandatory Welfare-to-Work Programs?', *Review of Economics*, 86: 156–179.
Mill, J. S. (1856) *A System of Logic* (New York: Harper & Brothers).
Nagel, E. (1961) *The Structure of Science: Problems in the Logic of Scientific Explanation* (New York: Harcourt, Brace, & World).
Neopolitan, R. (2004) *Learning Bayesian Networks* (Upper Saddle River, NJ: Prentice Hall).
Okruhlik, K. (1994) 'Gender and the Biological Sciences', *Canadian Journal of Philosophy*, 20: 21–42.
Paige, J. (1975) *Agrarian Revolution: Social Movements and Export Agriculture in the Underdeveloped World* (New York: Free Press).
Papineau, D. (1978) *For Science in the Social Sciences* (New York: St. Martin's Press).
Pearl, J. (2000) *Causality: Models, Reasoning, and Inference* (Cambridge: Cambridge University Press).

Pietroski, P. and G. Rey (1995) 'When Other Things Aren't Equal: Saving Ceteris Paribus Laws from Vacuity', *British Journal for the Philosophy of Science*, 46: 81–110.

Polhaus, G. (2002) 'Knowing Communities: an Investigation of Harding's Standpoint Epistemology', *Social Epistemology*, 13: 283–293.

Popper, K. (2002 [1957]) *The Poverty of Historicism* (New York: Routledge).

Rabinow, P. and W. Sullivan (1987) 'The Interpretive Turn', in P. Rabinow and W. Sullivan (eds), *Interpretive Social Science: a Second Look* (Berkeley: University of California Press), 1–30.

Rolin, K. (2006) 'The Bias Paradox in Feminist Standpoint Epistemology', *Episteme: a Journal of Social Epistemology*, 3: 125–136.

Roseberry, W. (1982) 'Balinese Cockfights and the Seduction of "Anthropology"', *Social Research*, 49: 1013–1028.

Roth, P. (1987) *Meaning and Method in the Social Sciences: a Case for Methodological Pluralism* (Ithaca: Cornell University Press).

Rothschild, E. (2001) *Economic Sentiment: Adam Smith, Condorcet, and the Enlightenment* (Cambridge, MA: Harvard University Press).

Schabas, M. (2005) *The Natural Origins of Economics* (Chicago: University of Chicago Press).

Scheines, R. (2005) 'The Similarity of Causal Inference in Experimental and Non-Experimental Studies', *Philosophy of Science*, 72: 927–940.

Schurz, G. (2002) 'Ceteris Paribus Laws: Classification and Deconstruction', *Erkenntnis*, 57: 75–96.

Searle, J. (1984) *Minds, Brains and Science* (Cambridge, MA: Harvard University Press).

Segal, D. (1999) 'A Response to Jones's Critique of Interpretive Social Science', *Philosophy of the Social Sciences*, 29: 306–309.

Solomon, M. (2001) *Social Empiricism* (Cambridge, MA: MIT Press).

Solomon, M. (2006) 'Norms of Epistemic Diversity', *Episteme: a Journal of Social Epistemology*, 3: 23–36.

Spirtes, P., C. Glymour, and R. Scheines (2000) *Causation, Prediction, and Search*, 2nd edn (Cambridge, MA: MIT Press).

Steel, D. (2008) *Across the Boundaries: Extrapolation in Biology and Social Science* (New York: Oxford University Press).

Steinmetz, G. (ed.) (2005) *The Politics of Method in the Human Sciences: Positivism and its Epistemological Others* (Durham, NC: Duke University Press).

Strevens, M. (2003) 'The Role of the Priority Rule in Science', *Journal of Philosophy*, 100(2), 55–79.

Taylor, C. (1971) 'Interpretation and the Sciences of Man', *Review of Metaphysics*, 25: 3–51.

Winch, P. (1958) *The Idea of a Social Science and its Relation to Philosophy*, 2nd edn (London: Routledge).

12
New Approaches to the Division of Cognitive Labor

Michael Weisberg

Scientists are not lone agents, cut off from the outside world, responding only to information generated in their own laboratories. Rather, they make decisions about what to investigate by integrating what they discover for themselves with what they learn from others. They also take into account external factors such as grants, prizes, and prestige. These sources of feedback lead scientists to coordinate and divide their resources among differing approaches to the research domain. This coordination seems to enhance the success of scientific communities, but this coordination is neither planned nor explicit. Philip Kitcher has called this fact about scientific communities *the division of cognitive labor*.

The division of cognitive labor is one of the most striking features of modern scientific communities and has been argued to be a key component in their epistemic success (Gerson, 2008; Hull, 1988; Kitcher, 1992, 1993; Rosenberger et al., n.d.; Solomon, 2001; Strevens 2003, 2006; Weisberg and Muldoon, forthcoming; Zollman 2007, forthcoming). However, relatively little is known about the connection between these cognitive and social facts about science and the epistemic productivity of scientific communities. Research questions concerning the division of cognitive labor are both empirical and theoretical. Empirically, we might ask: What is the actual division of labor in particular research communities? How much overlap and repetition of research is found in the most successful communities? What is the normal extent and pattern of communication among scientists? How do scientists make choices about dividing their cognitive labor? What is the connection between this division and the division of expertise documented by psychologists (Keil et al., 2008)? While philosophers might contribute to these questions, they tend to fall into the domains of psychology and sociology of science.

But there are theoretical questions concerning cognitive labor that are more easily addressed with philosophical methods: What different types of divisions of cognitive labor are possible? How effective are these divisions for achieving scientific goals? Are there tradeoffs among these divisions? What kinds of individual motivations can lead to these divisions? How do restrictions of information and resources affect these choices and the division of labor which is an outcome of the choices? What kinds of incentives or structural features might the scientific community adopt to achieve better divisions of cognitive labor?

These questions are at the heart of new work in philosophy of science that investigates the social structure of science. This work examines how the division of cognitive labor contributes to science's epistemic power, and how it can create epistemic and social problems. A truly interdisciplinary part of philosophy of science, this work draws together philosophers, historians, sociologists, psychologists, computer scientists, and mathematicians. In this chapter I will discuss three recent approaches to studying cognitive labor: Kitcher's and Strevens's *marginal contribution/reward* (MCR) approach, Zollman's and Grim's *epistemic networks* approach, and the *epistemic landscape* approach I developed with Ryan Muldoon.

1. Marginal contribution/reward (MCR) approach

The best-known approach for modeling cognitive labor has been championed by Philip Kitcher (1992, 1993) and Michael Strevens (2003, 2006). Kitcher and Strevens represent the division of cognitive labor as a community-level resource allocation problem. Imagine that the scientific community is trying to find the metabolic pathway of an important biological process and they can take several approaches to discovering this pathway. To maximize its chances of discovering the pathway in the minimum amount of time, the community needs to find a way to divide its most important resource, scientists, across these different projects. The optimal allocation will be the one that maximizes the probability of finding the pathway, or of finding it in the shortest amount of time or for the least cost.

While Kitcher and Strevens discuss how the community might calculate the optimum distribution of cognitive labor, the most interesting part of their analysis is conducted from the point of view of the individual scientist. Strictly speaking, they model a *representative agent*, but it is easier to think about the analysis from the point of view of a scientist newly entering the field. In their account, the scientist knows the current distribution of scientists to projects, as well as the *success function* for each project. Success functions represent the ability of the project to transform the cognitive resources of scientists into successful outcomes.

With the success function and the current distribution of cognitive labor, the model scientist can calculate its marginal contribution to the success of the different projects. In other words, she can figure out how much more probable a project's success would be if she were to join the project.

On the basis of such models, Kitcher and Strevens put forward an important argument: Classical epistemic norms will lead scientists to misallocate their cognitive labor. If a classically rational, truth-seeking agent followed the procedure above, then she would join the project with the highest probability of success. But this isn't always what the scientific community as a whole wants to see happen. Maximizing the chance of success might involve distributing scientists across projects, not just to the projects with the best chance of success.

Returning to our example, imagine that one of the ways of investigating the biochemical pathway had a high probability of success when a reasonable number of scientists works on the project. At the same time, there is another approach that has a relatively low probability of success, but this probability of success could be realized if a small number of scientists work on the project. In a case like this one, assuming diminishing marginal returns for increasing numbers of scientists, the community would be better off if a small number of scientists worked on the second approach. However, this would not happen if everyone followed the classical epistemic norm: 'take the approach most likely to lead to the truth'.

To explain this discrepancy between classical epistemic norms and real scientific communities, Kitcher and Strevens draw on analyses from the history and the sociology of science. Scientists, they argue, are often motivated by more than just epistemic factors. Prestige, credit, money, jealousy, greed, and so forth, are all factors that may affect how scientists choose to divide their labor.

Consider the simple case where scientists are motivated by credit alone. Assuming that the lion's share or all of the credit goes to the scientist who makes a discovery first, scientists will want to take into account both the probability of success of the project and the probability that they will be the first one to complete the project. The first consideration pushes scientists towards the project with the overall highest probability of success, but the second consideration pushes scientists towards projects that have fewer scientists working on them. This fact, Strevens (2003) has argued, explains why the scientific community has adopted the Priority Rule, the rule that whoever discovers something first gets all the credit.

This is only a sample of the utility of Kitcher's and Strevens's framework for studying cognitive labor, but it illustrates the kind of question that it can be most usefully applied to. When a fixed number of projects are available, and the incentives, success functions, and distribution of cognitive labor are known to all scientists, analytical results can be derived in this framework. Many of these results show that the scientific community is often right to hedge its bets by distributing cognitive labor. One way this can be encouraged is if the community organizes incentives so that self-interested individual scientists can lead the community as a whole to better fulfill its epistemic goals.

Productive as it has been, the MCR framework has been less successful in modeling situations of imperfect information and bounded rationality. For example, the models require that every agent know what every other agent is doing. The next approach to cognitive labor is specifically designed to investigate situations where this is not the case.

2. Epistemic network approach

A second approach to modeling cognitive labor focuses on how the social structure of science affects learning, confirmation, and the propagation of error. Famous laboratories such as the wartime Los Alamos nuclear weapons laboratory (Rhodes, 1987) and MIT RADAR laboratory (Galison, 1997) planned ways for scientists to continuously integrate their findings and share ideas with one another. Technological innovations such as the Internet and rapid forms of electronic publication make this possible on a much wider and geographically distributed scale. But is this high degree of collaboration an unambiguously good thing? Might such communication lead to the propagation and fixation of errors as well as knowledge?

The *epistemic network approach* investigates these questions by combining ideas from confirmation theory and network theory. Scientists attach probabilities to hypotheses using information that they discover for themselves and information that they learn from other members of their community. In new work by Grim (Rosenberger et al., n.d.) and Zollman (2007; forthcoming), lines of communication between scientists are represented by network graphs, such the ones seen in Figure 12.1. Each node of these graphs represents a scientist and each edge a communication channel. By altering the connectivity of the graph, from the minimally connected cycle to the maximally connected complete graph,

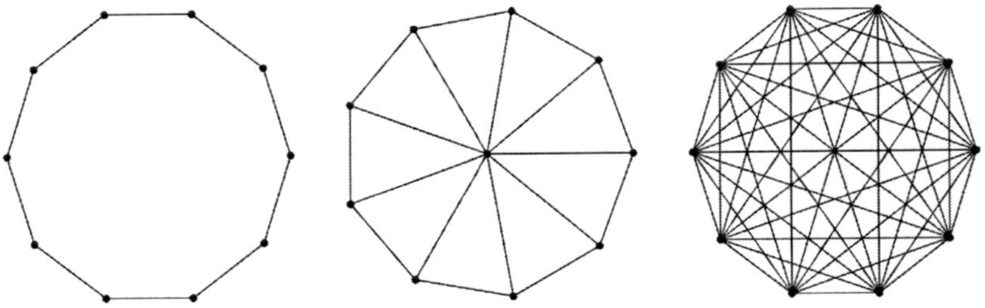

Figure 12.1 Zollman's three epistemic networks with 10 nodes. From left to right: the cycle, the wheel, and the complete network

Grim and Zollman have simulated different communication structures in science.

Zollman's recent work provides both an illustrative example of the potential of this work, and also a surprising and startling conclusion. He asks us to consider the following abstract scenario: scientists are trying to determine whether the world is in state φ_1 or φ_2. He likens this to trying to determine the causative agent for some disease. Are peptic ulcers, for example, caused by hyperacidity or bacteria? In standard Bayesian fashion, every scientist has a prior probability on the hypotheses that φ_1 is true and that φ_2 is true, where no scientist attaches probability 1 or 0 to either hypothesis and where these two hypotheses are mutually exclusive.

Now imagine that data start coming in by experiment. As each scientist begins experimenting, she receives information about the world. This information is drawn from a distribution centered around the objectively true answer, taking into account the fact that experiments produce noisy data. Scientists update their beliefs according to the data that they generate, as well as the data generated by the peers to whom they are connected in a social network.

Zollman took this basic idea and applied it to network models suitable for investigation by simulation. He iterated the scenario described above until the community converge either on the correct answer or on another state where the population is no longer able to take in new information. The scenario was studied with randomized initial priors and repeated 10,000 times per condition.

Although there are many subtleties to Zollman's results, two especially interesting conclusions can be drawn from this work. First, scientists connected in a cycle converged to the truth more often than scientists connected in a wheel or in a fully connected graph. This suggests the unintuitive conclusion that careful limiting of information available to scientists may have certain advantages. Or to put the point more bluntly, less well-informed scientists might have an advantage over more well-informed ones, if the goal is to minimize error. The second result, however, suggests an advantage of the more highly connected communities. When these communities converge to the truth, they do so more rapidly. For the ten-scientist communities Zollman studied, those on complete networks converged about five times faster than those in the cycle network.

The epistemic network approach promises to be extremely helpful in studying divisions of cognitive labor that result in limited information to individual scientists. Grim and Zollman have primarily used this method to address issues of confirmation, but one could also imagine studies of project choice in this framework as well. One could start with the MCR framework, but then include network structure to explicitly limit the information available to scientists in making a choice about how to divide their cognitive labor.

Despite its advantages, the epistemic networks approach still requires a lot of initial fixity. The hypotheses under consideration as well as the network structure need to be fixed ahead of time and do not change. In addition, the models are homogeneous with respect to scientists. Every scientist is epistemically equivalent to all the others, the only difference being in connectivity. The final approach to cognitive labor allows us to have heterogeneous populations and allows the space of projects and hypotheses to change 'on-line', a more accurate reflection of the way science proceeds.

3. Science on an epistemic landscape

A third approach to modeling the division of cognitive labor is the *epistemic landscape* approach, which I have developed along with Ryan Muldoon. This approach incorporates aspects of MCR and epistemic networks, but envisions scientific research as being more like foraging than either microeconomic optimization or Bayesian reasoning. Like epistemic network models, this approach severely limits the information available to scientists. The models focus, however, on MCR-like questions about how scientists choose to divide their cognitive labor. The approach starts with agents of very limited knowledge and rationality and tests how interesting dynamics can arise from the bottom up.

Like the other approaches to cognitive labor considered in this chapter, the epistemic landscapes approach begins from the individual scientist's point of view. We start by considering the epistemic situation of individual scientists – what information they have, what their motivations are, and so forth. We then try to work out what rules scientists might follow when making decisions about which research projects to pursue and implement these rules in the models.

Let's start from a concrete example. Consider a psychologist interested in how children's understanding of pretense develops. Given that this is a new area, there are few guidelines about what specific projects she might choose to work on. She does know that another lab has demonstrated the basic developmental milestones of spontaneous pretend play, but this only constrains her next move very modestly. Assuming that she is early in her career, a reasonable first move is to start off by replicating the previous finding or by choosing a very similar project. Then, as she makes small changes from this initial approach, she can determine the extent to which the resulting findings are epistemically significant.

Although this sounds straightforward in the abstract, there are actually many degrees of freedom making the choice of the next project non-trivial. Even if the scientist starts by investigating the developmental trajectory of pretend play and wants to 'take the next step', what direction should she take it in? This is where computational models can help us start to get a handle on the process of project choice.

Epistemic landscape

Epistemic landscape models begin by postulating a set of *approaches*, narrow specifications of how a research topic is investigated. Approaches specify the research questions being investigated, the instruments and techniques used to gather data, the methods used to analyze the data, and the background theories used to interpret the data.

In the psychology example described above, different approaches might involve investigating these phenomena in children of different ages, looking at the differences between individual and group play, considering how children play with peers vs. adults, etc. Within these classes of approaches would be the specifics of the research method, including where the population was drawn from, direct manipulation versus observational approaches, the props used to initiate pretend play, and the like. Finally, the same techniques aimed at the same questions may yield different results in light of the background theories used to interpret the data. Approaches are narrowly individuated in all of these respects and represented spatially in our models.

Scientists make strategic choices when they choose or modify their approaches. In particular, they want to choose approaches that generate results of equal or greater significance to their current approach. So we need a way of representing the connection between approaches and the significance of the knowledge generated by using that approach.

For present purposes, let's confine ourselves to what Kitcher (2001) has called the *epistemic significance* of scientific knowledge. This is the purely scientific value of a result that the community agrees on. Financial and other pragmatic values are excluded for present purposes. Let's further suppose that all of our scientists are sufficiently talented that adopting an approach will always yield whatever significant truths could be discovered by employing that approach.

With these assumptions, we can construct an *epistemic landscape*. The dimensions of this landscape correspond to aspects of approaches, along with an additional dimension of epistemic significance. Each point is thus an approach with a corresponding degree of epistemic significance.

Epistemic landscapes are the scientist-independent part of this representational framework. Here, only objective information about the world is represented. It takes the form: If such and such an investigation (narrowly construed) is performed by a competent scientist, then such and such results will obtain. All of this information is meant to be independent of any particular scientist.

In order to make the landscape tractable to simulation, we need to take three further, pragmatic steps. First, we discretize the landscape so that 'patches' instead of points correspond to approaches. Second, we reduce the dimensionality of the landscape from the hundreds of dimensions in a realistic landscape to three: two for approach and one for significance. This reduction can be thought of as either 'bundling' aspects of approaches together, or more plausibly as a subspace of the larger approach space. Finally, we wrap the landscape on a torus so that it has no edges. When we do this, we get a landscape much like the one depicted in Figure 12.2, which is the landscape used for the tests described in this chapter.

Scientist agents

The second half of the framework brings us back to the scientist's point of view. In the real world, a scientist knows what approaches she has taken in the past and the success of these approaches. She also knows what approaches *some* of her colleagues have taken and how successful they were. What a scientist doesn't know is the entire topography of the epistemic landscape. She can't be certain how significant untried approaches will prove to be. All she can do is to make inferences about

Figure 12.2 The example epistemic landscape used in the simulations described in this chapter

these approaches and then try them out. These facts are all reflected in the epistemic landscape framework.

In epistemic landscape models, a scientist's current approach is represented spatially: scientists' locations on the landscape correspond to their current approaches. The model scientist maintains some memory about where she has been before, so that she can determine later if she is going in a plausible direction. Further, by adopting a particular approach, scientist agents are able to determine the significance associated with their current approach. This corresponds to a real-world scientist conducting successful research and determining how significant her results actually are.

Since epistemic landscapes are supposed to be models of the social structure of science, agents need to have information about the activities and progress of other scientists. Real scientists devote time to reading the literature, attending conferences, and communicating with colleagues, learning what has been tried and what has been successful. In the models described below, this will be represented in a simplified manner, which roughly corresponds to reading the scientific literature or going to conferences. Agents will be able to see a limited range of other approaches, determine whether these approaches have been tried and if so, discover the degree of significance associated with these approaches.

The models Muldoon and I (forthcoming) have developed focus on approach choice, much like MCR models. Once our model scientists can determine the significance of the truths discovered with their current approach and their neighbors' approaches, they must decide what approach to try next. A major advantage of epistemic landscape models is that these decisions can be made differently by different agents; each can have its own strategy. Our preliminary analyses have investigated three types of agents with different strategies: *controls*, *mavericks*, and *followers*.

Controls

We begin by considering model scientists with extremely limited knowledge: they only know about what approaches they have taken in the past and the significance of the truths generated by taking these approaches. These scientists are unaware of or uninterested in the rest of the community. Their only interest is in surveying the epistemic landscape in order to find areas of high significance. In addition, the scientists will engage in some degree of experimentation: behavior that deviates from their current research trajectory. This will keep the agents from getting marooned in zero significance areas. Because such scientists do not take others' discoveries into account, they will serve as our baseline case for investigating the division of cognitive labor.

In our epistemic landscape models, scientists choose their approaches by employing strategies. The controls deploy the following, very simple strategy:

CONTROL

1. Move forward one patch.
2. Ask: Is the patch I am investigating more significant than my previous patch?
   - If Yes: Move forward one patch.
   - If No: Ask: Is its significance equal to the previous patch?
     - If Yes: With probability *e*, move forward one patch with a random heading. Otherwise, do not move.
     - If No: Move back to the previous patch. Set a new random heading.
3. With probability *e*, change to a random heading and move forward.

As discussed above, this strategy restricts the agents' knowledge to what they can detect themselves about the significance of a patch. They will not even notice if another agent is currently on the same patch. The rule ensures that their movement will always be in the direction of increasing significance if they are on a gradient. If they get stuck in a low significance area, the experimentation probability will eventually cause them to move in a new direction and find a more significant area. Scientists following this search pattern are guaranteed to find local significance maxima in finite time. However, the experimentation procedure is not strong enough to knock scientists off of local maxima, unless the peak is constrained to a single patch. This means that the controls are guaranteed to find at least one approach of peak significance, but not necessarily both.

To assess the research potential of a community of controls, Muldoon and I asked the following questions:

1. How fast does a community of controls find the two peaks of the epistemic landscape? How does this scale up as the number of scientists increases?
2. How much epistemic progress does the community of scientists make? How does this scale up as the number of scientists increases?

These questions were addressed by repeatedly employing the landscape shown in Figure 12.2, but varying the initial positions and numbers of scientists. With respect to the first question, we found that ten controls found both peaks 95 percent of the time, but that the time it took to find these peaks varied from about 550 to about 43,000 cycles of the model. This variance seems to be explained solely as a function of the initial distribution of scientists to approaches, which were randomly assigned, although agents were restricted to low significance areas. Increasing the number of agents ensured that both peaks were found (as happens 100 percent of the time in our simulations with 20 agents) and that the mean time to finding these peaks was rapidly diminished with increasing numbers of controls.

While it is obviously important for the scientific community to find the epistemic landscape's peaks, much important research also happens using non-maximally significant approaches. To investigate this, let *epistemic progress* be the fraction of patches with significance greater than zero that have been visited by the community of scientists.

Employing the same epistemic landscape as before, we discovered a linear relationship between the number of controls and the average epistemic progress of the community. For any fixed number of model cycles, increasing the number of controls gives a linear increase in the average epistemic progress of the community. Fixing the number of controls and increasing the number of model cycles also results in a linear increase of epistemic progress. In more recent work, my students and I have shown that this result is robust across many changes to the epistemic landscape (Werner et al., 2008). Specifically, changing the number, location, and size of the hills and peaks do not change these results in any appreciable, qualitative way.

Despite their modest success, a community of controls is neither very effective nor very efficient. Large populations of controls can achieve high degrees of epistemic progress, but it takes a considerable amount of time for this to happen. An obvious reason for this is that controls cannot learn from one another or take into account what other scientists are doing when they plan their next moves. Communities of controls do not divide cognitive labor; each member of the community is acting as if it were the only member.

Getting social: followers and mavericks

In the last section, I discussed the behavior of a classically rational agent: one who simply tries to find approaches yielding truths of the highest significance. This involves no division of cognitive labor at all. Each agent acts as if she was the only one. Understanding the behavior of these agents first provides a baseline against which we can now compare the behavior of scientists who do divide their cognitive labor.

Followers divide their cognitive labor by reasoning that the best way to find more significant truths about the world is to find the approach which has yielded the highest significance so far, and move in that direction. This is simulated in several steps. At the beginning of each cycle of the model, followers examine the patches in their Moore neighborhood, the 8 patches immediately adjacent to the one on which they are currently located. The agents then move to the previously explored approach of maximum significance in their Moore neighborhood, if such an approach is available. More specifically, followers execute the following decision procedure:

FOLLOWER

Ask: Have any of the approaches in my Moore neighborhood been investigated?

- If Yes: Ask: Is the significance of any of the investigated approaches greater than the significance of my current approach?
  - If Yes: Move towards the approach of greater significance. If there is a tie, pick randomly between them.
  - If No: If there is an unvisited approach in the Moore neighborhood, move to it, otherwise, stop.

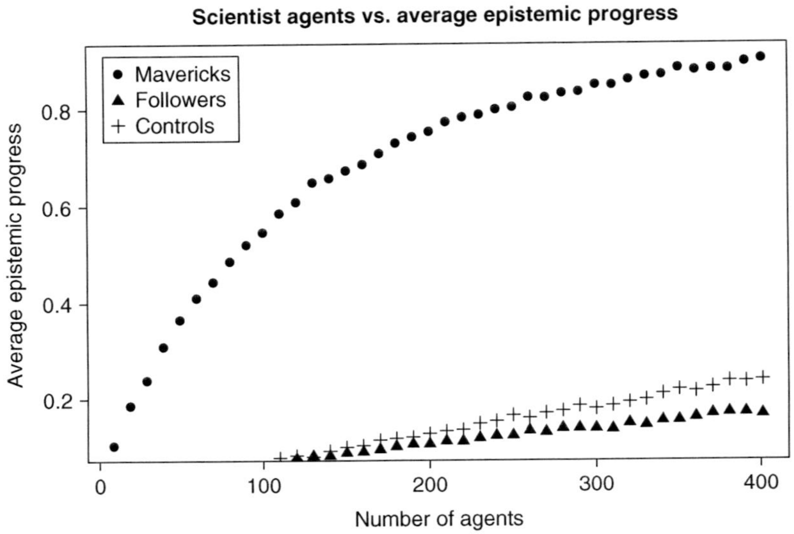

Figure 12.3 Comparison of the epistemic progress of controls, followers, and mavericks (Controls and mavericks measured after 200 cycles, followers after 1,000)

If No: Choose a new approach in the Moore neighborhood at random.

A parallel set of simulations to the ones described above reveals that followers are surprisingly bad at finding peaks and making epistemic progress. With 200 followers, a single peak (approach of maximum significance) was found 60 percent of the time, with both approaches being found only 12 percent of the time. In addition, as we can see in Figure 12.3, the average epistemic progress of a community of 400 followers is only 0.17 on our diagnostic landscape, which means that only 17 percent of the significant approaches were discovered. This contrasts poorly with communities of controls. It took fewer than 300 controls to reach the same significance achieved by 400 followers.

These results suggest that the high degree of coordination and learning exhibited by followers is not an optimal strategy, at least when it is shared by the entire community. This result might lead one to think that learning from others inevitably leads to bad epistemic outcomes – indeed, this is one lesson we might learn from Zollman's models. But is the problem coordination with other agents, or the way that followers coordinate? The third strategy will let us investigate this question.

Like followers, *mavericks* pay attention to what others are doing, but they use this information differently. Instead of moving towards approaches yielding high significance, mavericks move away from explored territory.

At the beginning of each cycle of the model, mavericks examine the patches in their Moore neighborhood and execute the following decision procedure:

MAVERICK

Ask: Is my current approach yielding equal or greater significance than my previous approach?
- If Yes: Ask: Are any of the patches in my Moore neighborhood unvisited?
  - If Yes: Move towards the unvisited patch. If there are multiple unvisited patches, pick randomly between them.
  - If No: If any of the patches in my neighborhood have a higher significance value, go towards one of them, otherwise stop.
- If No: Go back one patch and set a new random heading.

Mavericks' performance is vastly superior to that of controls and followers. The maximally significant peaks are always found, even with very small populations. As with the controls and followers, we examined populations of 10 to 400 mavericks in increments of 10. With 100 mavericks, the community achieves 0.55 epistemic progress after 200 model cycles. With 400 mavericks, they achieve epistemic progress of 0.90 in the same time. Mavericks are thus extremely efficient at finding peaks and, due to the methods they use to find the peaks, they also make excellent and rapid epistemic progress. Figure 12.3 makes the contrast between mavericks and followers clear and shows that, while it may be useful for scientists to take into account what others have discovered, this information can be detrimental to progress when taken into account in the wrong way.

Polymorphic populations

It seems highly unlikely that a real scientific community would be composed exclusively of mavericks or followers. Strategies for scientific inquiry vary between scientists and within careers. Some scientists are more individualistic and maverick-like; others follow trends and are follower-like. So what happens in the epistemic landscape model when populations are mixed?

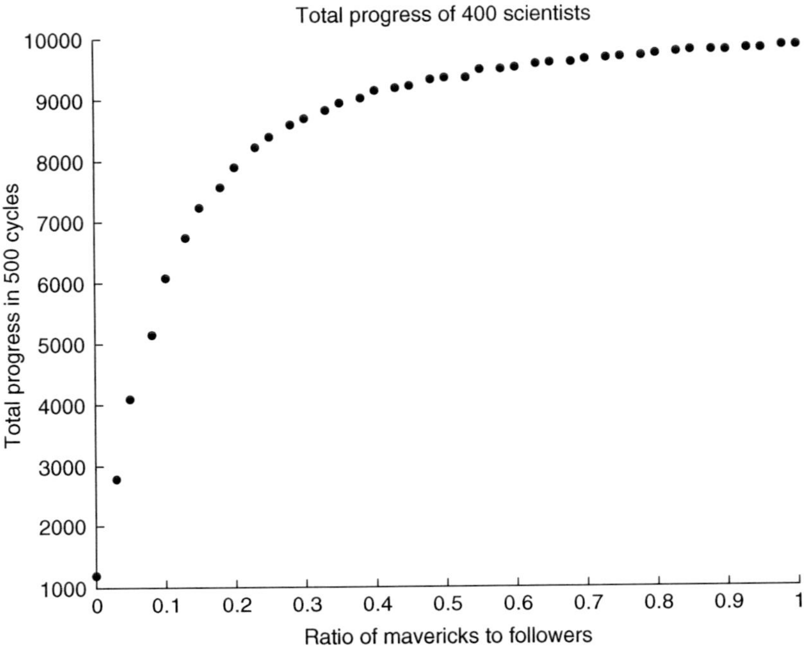

Figure 12.4 Average number of approaches investigated by mixed populations of followers and mavericks after 500 model cycles (The landscape consists of 10,201 approaches total. The population size is held fixed at 400 scientists, but the ratio of mavericks to followers is varied from 0 mavericks to 400 mavericks)

An initial result for mixed populations is that the substitution of a *single* maverick for a follower results in a statistically significant increase of epistemic progress made by a community composed of followers. This increase of epistemic significance, however, is small. To investigate this effect further, Muldoon and I studied populations of scientist-agents with fixed size, but varied the ratio of mavericks to followers. We found that even small numbers of mavericks mixed in with followers causes the population to make considerably more epistemic progress.

To better understand why this happened, we adopted a second diagnostic measure, the *total progress* of the community. Total progress corresponds to the fraction of approaches investigated, whether these approaches are significant or not. This measure allows us to see how much total activity is being performed by the scientific community.

Figure 12.4 summarizes the results, and we can see that the initial addition of mavericks (ratios of 0.02–0.10) causes rapid tripling then quadrupling of the number of approaches investigated. Further small increases in the number of mavericks (0.10–0.40) take the population to around the 90 percent mark for the number of approaches explored in 500 cycles. This massive increase in total progress is primarily a result of the increased stimulation of followers by mavericks. As mavericks pass through a region where followers are located, they effectively 'unlock' approaches for investigation that the followers would have otherwise avoided. This means that in addition to their own direct effects on epistemic and total progress, mavericks make a very important indirect contribution to epistemic and total progress by stimulating followers.

Despite the fact that mavericks stimulate the followers to make considerable epistemic and total progress, pure populations of mavericks still perform better than equivalently sized mixed populations. Scientist for scientist, mavericks make more epistemic progress and more total progress than populations of followers. Yet as I mentioned above, scientific communities are typically polymorphic in strategy. Why might this be?

One possibility is that the maverick strategy is more costly both to individuals and to the community. In real communities, this could be due to the strategy's extreme anti-conservatism. Mavericks cannot learn from their neighbors, or even borrow their neighbor's techniques, equipment, background research and the like. Unless one had a very large research budget consisting of lots of money, supplies, and helpers, it would be professionally, institutionally, and personally very costly to be a maverick.

If it is more costly to be a maverick, then when resources are finite, optimum research communities should be composed primarily of followers, but with some mavericks included. In a series of new simulations, I have investigated this possibility by adding two additional aspects to the model: research resources and research cost.

I start off with a fixed pool of available resource units. Research costs are represented as the consumption of units of this resource per agent per cycle of the model. To investigate situations where mavericks are more costly than followers, the followers consume one unit per cycle, while mavericks consume ten units per cycle. If polymorphism in real populations of scientists is explained by research costs, then imposing these restrictions ought to reveal a tradeoff: in a fixed population of scientists with modest resources, there will be some optimum balance of mavericks and followers, mostly consisting of followers.

Investigation of the new, resource-limited models is ongoing, but two preliminary results are worth noting. The first result is that mavericks are always better at making epistemic progress than followers. My students and I expected that imposing modest resource limits and making each maverick more expensive than a follower would result in a tradeoff where a mixed population of mavericks and followers made maximal epistemic progress; we were incorrect. Mavericks are so much better than the followers that in most scenarios, pure populations of mavericks are better than mixed populations, despite the fact that they have many fewer cycles to investigate the landscape.

Only when we imposed severe restrictions – ones that caused a pure population of mavericks to only run for about 25 cycles – did we start to see the tradeoff (Figure 12.5). Even here, the optimum balance of mavericks and followers is shifted much further in the direction of mavericks than we expected. While these models are still relatively simple, these results suggest that higher research costs cannot be the whole explanation for polymorphism in research communities.

Epistemic landscape models' flexibility for representing cognitive labor should be evident, even from this brief presentation. To date, however, they still lack some of the positive features of the other approaches. Unlike the MCR approach, epistemic landscape models do not provide an easy way of determining the *community-level* optimum distribution of labor. Different distributions need to be tried by simulation, upon specification of the measure of interest. Similarly, landscape models have not yet represented community structure other than Moore neighborhoods of approaches. It is easy to generalize the Moore neighborhood concept to bigger areas, but this is still a neighborhood of approaches, not of agents.

4. Connecting the three approaches

The models described in this chapter are all extremely simple. Each one of them can be extended in myriad ways, and some of their key results might be tested by looking at sociological and psychological data. Rather than describe these possible extensions and empirical tests, I want to close with a few comments about how the central ideas of the three modeling approaches might inform one another in future work.

One way of uniting the approaches is to take conclusions generated by MCR models, and investigate the conditions under which they could arise using network and landscape models. For example, say we wanted to discover the conditions under which the scientific community would adopt the Priority Rule. An agent-based model, similar to the ones used in the epistemic landscape approach, could be used to relax some

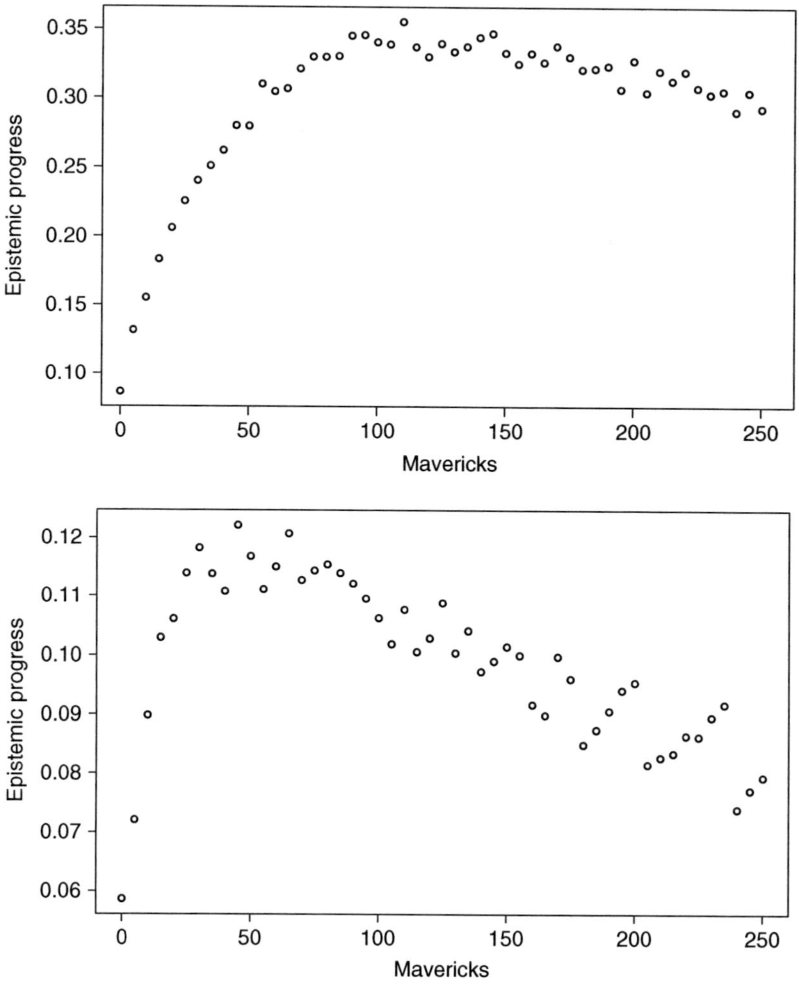

Figure 12.5 The effect of resource restrictions on epistemic progress (Above, resources are restricted to a total of 75,000 agent-moves, where each maverick move costs ten times more than a follower moves. Below, only 25,000 agent-moves are allowed. In both cases, the populations have a total of 250 scientists, divided between followers and mavericks)

of Strevens's key assumptions, seeing under what conditions the Priority Rule produces the optimal distribution of labor. Another model might incorporate an evolutionary dynamic (with mentors 'giving birth' to students) and show under what conditions the Priority Rule would spontaneously arise. The approaches might also make contact when additional sources of value or significance (practical, monetary, etc.) are included in the landscape. Here one might investigate the interaction of reward schemes and strategies with different epistemic landscapes.

A second potential connection between the three approaches would be to incorporate epistemic networks into MCR and landscape models. Their incorporation into MCR models would be one way to introduce bounded rationality into models that otherwise assume that representative agents have access to all relevant information. One might also imagine using epistemic networks in landscape models in order to create more realistic versions of the follower and maverick strategies. The work of some scientists will be particularly salient to others, because of fame, skill, friendship, or other factors. It is reasonable to assume that followers only follow some scientists, and not necessarily just the ones employing nearby approaches. Similarly, mavericks may be comfortable following some very important leaders in their field, but otherwise they want to avoid what most other scientists are doing.

Finally, ideas from the landscapes approach might be fruitfully adopted in epistemic network models. While the connectivity in network models can be varied, the actual epistemic strategy of each scientist is homogeneous and based on simple Bayesian reasoning. As the landscape approach shows, interesting results arise when strategies are varied and populations are polymorphic. Similar tests could be deployed using epistemic networks, or the basic machinery of epistemic networks could be employed in landscape models.

References

Hull, D. (1988) *Science as a Process: an Evolutionary Account of the Social and Conceptual Development of Science* (Chicago: University of Chicago Press).

Keil, F. C., C. Stein, L. Webb, V. D. Billings, and L. Rozenblit (2008) 'Discerning the Division of Cognitive Labor: an Emerging Understanding of How Knowledge is Clustered in Other Minds', *Cognitive Science*, 32(2): 259–300.

Kitcher, P. (1992) 'The Division of Cognitive Labor', *Philosophy of Science*, 87: 5–22.

Kitcher, P. (1993) *The Advancement of Science* (New York: Oxford University Press).

Kitcher, P. (2001) *Science, Truth, and Democracy* (New York: Oxford University Press).

Galison, P. L. (1997) *Image and Logic: a Material Culture of Microphysics* (Chicago: University of Chicago Press).

Gerson, E. M. (2008) 'Reach, Bracket, and the Limits of Rationalized Coordination: Some Challenges for CSCW', in M. S. Ackerman, C. Halverson, T. Erickson, and W. A. Kellogg (eds), *Resources, Co-evolution, and Artifacts: Theory in CSCW* (New York: Springer-Verlag), 193–220.

Rhodes, R. (1987) *The Making of the Atomic Bomb* (New York: Simon & Schuster).

Rosenberger, R., P. Grim, B. Anderson, A. Rosenfeld, and R. Eason (ms, n.d.) 'Science in Simulation: Structure and Evaluation'.
Solomon, M. (2001) *Social Empiricism* (Cambridge, MA: MIT Press).
Strevens, M. (2003) 'The Role of the Priority Rule in Science', *Journal of Philosophy*, 100(2): 55–79.
Strevens, M. (2006) 'The Role of the Matthew Effect in Science', *Studies in the History and Philosophy of Science*, 37: 159–170.
Weisberg, M. and R. (forthcoming) 'Epistemic Landscapes and the Division of Cognitive Labor', *Philosophy of Science*.
Werner, M., B. Naecker, R. Muldoon, and M. Weisberg (2008) 'Robustness Tests on Epistemic Landscapes' (http://philsci-archive.pitt.edu).
Zollman, K. J. S. (2007) 'The Communication Structure of Epistemic Communities', *Philosophy of Science*, 74(5): 574–587.
Zollman, K. J. S. (forthcoming) 'Social Structure and the Effects of Conformity', *Synthese*.

Index